The
Science
of
Gardening

The Science of Gardening

Discover how your garden really grows

DR STUART FARRIMOND

CONTENTS

INTRODUCTION

It has taken me forty years to discover the joy of gardening. Not since sowing a handful of cress seeds into a compost-filled plastic cup at the age of six had I felt the thrill of seeing perfectly formed green life emerge from brown dirt. But my adult hands, softened from years of tapped keys and pushed pencils, had once again made this miracle happen on a kitchen windowsill. It didn't matter that only three of the thirty Brussels sprout seeds made it and that their crop was tiny – a gate had been opened in my heart and mind into a world I was keen to explore.

Gardening is the perfect antidote to doom-scrolling through today's news, it reconnects us with the perpetual cycle of life, death, and renewal of which we are all a part. In fact, I can think of no other pursuit that offers more. Whether you have a balcony or smallholding, growing plants allows you to be a designer, sculptor, artist, wonder-struck child, and inquisitive scientist with ever more to learn.

And yet for something as beautifully simple as sowing, planting, and watering, we humans have made gardening terribly complicated. Trained as a medical doctor, I know only too well how to bamboozle others with technical terms – and now as a gardener I was being befuddled by a forest of gobbledygook and strange rituals. What is a perennial? What on earth is mulching? Why mustn't I water plants in the middle of the day? Our adult egos make us embarrassed to ask. But, old hand or rookie, I'm willing to wager that you've been confused by Latin names, a mysterious term, or perhaps wondered whether you really do need to put "crocks" in the bottom of plant pots.

Most gardening books and websites are missing a trick – they explain the "hows" of growing and plant care without even touching on the "whys". As a man of science whose passion is to demystify and debunk, I have taken delight in using science and the latest research to answer key gardening questions and reveal why some age-old practices belong in the past.

In the pages that follow I will reveal to you the science behind the "how" when caring for plants, and hope that it informs and inspires your gardening in the same way that it has mine.

WONDER OF GARDENING

HOW AMAZING ARE PLANTS?

Here are just a few of the more remarkable ways in which plants have evolved to survive and thrive, including making their own food, duping animals into transporting their pollen, and regenerating parts that have been eaten.

───────────

Plants give us every breath of oxygen that fills our lungs, make their food out of thin air in a process called photosynthesis, and are the original source of every mouthful we eat. They are the lynchpin of all life on earth and the ultimate survivors, managing to thrive in even the harshest conditions in practically every corner of the globe. Trees can weigh in at over 2,000 tonnes and can live to be over 5,000 years old. Some Great Basin bristlecone pine trees (*Pinus longaeva*) were alive even before Egypt's pyramids were built.

AT HOME IN HARSH CLIMATES

Few organisms can survive -50°C (-58°F) temperatures, yet plants have found a way. Moss campion (*Silene acaulis*) is a type of "cushion plant" that hugs the ground as a squat dome to escape the worst of the wind. It happily survives on frozen mountainsides under a blanket of snow, while biological antifreeze in its sap stops it freezing.

In the parched deserts of southwest Africa, the seeds of alien-like *Welwitschia* plants can wait for centuries until rainfall provides the right conditions for germination. In the Atacama desert, where years can pass without rain, desert moss (*Syntrichia caninervis*) stays alive by sucking moisture directly out of passing fog and mist. "Air plants" (*Tillandsia*) need no soil whatsoever, but use their wiry roots to cling to rocks, branches or cliff faces, pulling moisture directly from the air and harvesting nutrients by catching passing dust in their tiny leaf hairs (called trichomes). Unsurprisingly, *Tillandsia* plants tend to be very slow growing.

Extraordinary adaptations
Plants have evolved countless ingenious ways to survive and flourish in every habitat, from lush rainforests, to inhospitable deserts and mountainsides.

HURA CREPITANS
A TRUNK STUDDED WITH VICIOUS SPIKES IS JUST ONE OF THIS TREE'S DEFENCES

DWARF MISTLETOES
FIRE THEIR SEEDS LIKE BULLETS INTO NEIGHBOURING TREES' BARK

DISPERSAL AND DEFENCE

Plants have overcome their inability to move in astonishing ways. Large, bauble-shaped *Brunsvigia* is a type of tumbleweed whose flowerhead dries up, amputates itself, and then rolls off in the breeze, dropping its seed. Javan cucumbers (*Alsomitra macrocarpa*) produce seeds shaped like gliders that, launched from their high vines, will soar for hundreds of metres. Dwarf mistletoes (*Arceuthobium*), which grow as parasites on trees, expand their range by expelling glue-coated seeds at 60mph (97kph). As their fruits ripen, the flesh heats up rapidly (called thermogenesis) until the fruit ruptures and fires out seeds at high speed.

Living in a world of hungry creatures means plants have developed some sophisticated defence systems to avoid being eaten. Cacti have evolved to defend their water-filled bodies from attackers by turning their leaves into sharp spikes, while succulents known as "living stones" (*Lithops*) hide in plain site from thirsty animals by looking just like pebbles. The deadly "dynamite tree" (*Hura crepitans*) has lethal sap that is used in poison darts, and spreads its seeds with a bang. The tree's innocent-looking fruit is a grenade primed to explode. As it dries, tension builds up in the skin until eventually, with a tiny tap, the whole fruit breaks apart violently – each seed-loaded segment rocketing outwards at up to 156mph (251kph). Seeds are dispersed, and would-be predators get a nasty shock!

WELWITSCHIA
GROW SLOWLY IN HARSH DESERT CONDITIONS, BUT CAN LIVE FOR OVER 1,500 YEARS

JAVAN CUCUMBERS
MELON-LIKE FRUIT RELEASE GLIDING SEEDS WITH PAPER-THIN, 14CM (6IN)-WIDE WINGS

REBUTIA
SHARP SPINES DETER PREDATORS, BUT ALSO HELP PLANTS REGULATE TEMPERATURE AND WATER LOSS

BRUNSVIGIA
A SOUTH AFRICAN TUMBLEWEED THAT DISPERSES SEEDS AS IT'S BLOWN ALONG BY THE WIND

DO PLANTS HAVE INTELLIGENCE?

We don't think of plants as having intelligence – they have no brain, after all. But how is it then that our leafy friends are capable of many things that we usually think only creatures with intelligence can do?

Believing that plants can hear our comforting words is sadly wishful thinking (see p.208). Many of our ideas that plants have human-like smarts originated from the bestselling 1975 book, *The Secret Life of Plants*, in which dubious experiments supposedly proved that plants can read minds and become distressed at the sound of an egg cracking. These claims are fairy tales, but plants' real abilities are almost as extraordinary.

We are blind to much of what plants do simply because they live on a different timescale to us. A timelapse film of a plant exploring its surroundings, might almost be enough to convince you it is a thinking animal: tendrils from climbing plants methodically grope around for a surface to cling to;

growing sunflower seedlings swirl and dance with one another while tracking the sun; plant roots probe and feel the soil like fingers.

SENSING THEIR SURROUNDINGS

Nearly 150 years ago, Charles Darwin became convinced that root tips act as if they carry a "brain-like organ" and make decisions. Long filed as a silly idea, research has since revealed that root tips have at least 15 senses: sniffing air, tasting soil, sensing light and gravity, and using touch to avoid obstacles.

Plants have no eyes, yet they can "see" light in different colours thanks to microscopic light sensors called photoreceptors. These allow them to grow towards light, and even twist to follow the sun

Phototropism

Plants grow towards a light source (an ability called phototropism) to help them avoid the shade cast by neighbours and maximize the sunlight strking their leaves to power photosynthesis. Photoreceptor proteins sensitive to blue light wavelengths in sunlight trigger this response.

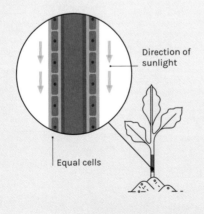

Direction of sunlight

Equal cells

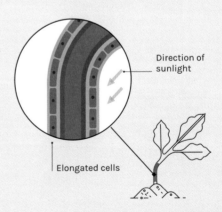

Direction of sunlight

Elongated cells

LIGHT FROM ABOVE

Straight stems and upright growth occur when sunlight strikes plants from above, bathing the leaves and stems evenly in blue light.

LIGHT FROM SIDE

Plants grow towards light because shaded cells receiving less blue light elongate faster than those in the sun, causing the stem to bend.

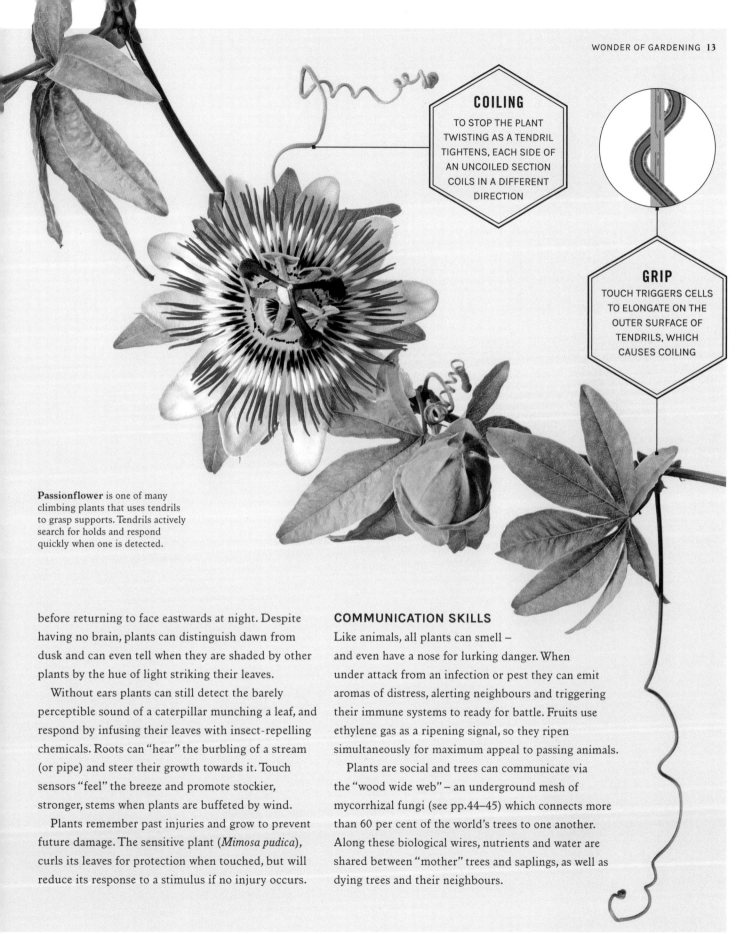

COILING

TO STOP THE PLANT TWISTING AS A TENDRIL TIGHTENS, EACH SIDE OF AN UNCOILED SECTION COILS IN A DIFFERENT DIRECTION

GRIP

TOUCH TRIGGERS CELLS TO ELONGATE ON THE OUTER SURFACE OF TENDRILS, WHICH CAUSES COILING

Passionflower is one of many climbing plants that uses tendrils to grasp supports. Tendrils actively search for holds and respond quickly when one is detected.

before returning to face eastwards at night. Despite having no brain, plants can distinguish dawn from dusk and can even tell when they are shaded by other plants by the hue of light striking their leaves.

Without ears plants can still detect the barely perceptible sound of a caterpillar munching a leaf, and respond by infusing their leaves with insect-repelling chemicals. Roots can "hear" the burbling of a stream (or pipe) and steer their growth towards it. Touch sensors "feel" the breeze and promote stockier, stronger, stems when plants are buffeted by wind.

Plants remember past injuries and grow to prevent future damage. The sensitive plant (*Mimosa pudica*), curls its leaves for protection when touched, but will reduce its response to a stimulus if no injury occurs.

COMMUNICATION SKILLS

Like animals, all plants can smell – and even have a nose for lurking danger. When under attack from an infection or pest they can emit aromas of distress, alerting neighbours and triggering their immune systems to ready for battle. Fruits use ethylene gas as a ripening signal, so they ripen simultaneously for maximum appeal to passing animals.

Plants are social and trees can communicate via the "wood wide web" – an underground mesh of mycorrhizal fungi (see pp.44–45) which connects more than 60 per cent of the world's trees to one another. Along these biological wires, nutrients and water are shared between "mother" trees and saplings, as well as dying trees and their neighbours.

IS GARDENING GOOD FOR ME?

Medics have long been taught to place their trust almost exclusively in medicines and scalpels. But times are changing, and "soft" treatments, such as walks in the park, time by the sea, or a prescription of horticultural therapy, are gaining ground.

———————

In a fast-moving world of quick fixes, gardening puts us back into the pace of life from which we evolved. Plants grow and die over days, weeks, and seasons; a rate similar to that at which our minds learn, our bodies grow, and our wounds heal.

GROWTH AND HEALING

For those suffering or recovering from psychological trauma, the garden is a calm, protected space away from the stresses of life. Psychiatrists and psychologists have found that gardening offers a safe place to learn how to care for something outside of ourselves when we have suffered past hurts and traumas. Some plants can sting or have thorns, but they are never angry and will not bite back.

Working in the garden provides a valuable opportunity for exercise – burning between 210 and 420 calories per hour, about the same as yoga or playing badminton – and cuts risk of heart attack, stroke, and diabetes, helps avoid weight gain, improves self-esteem, and protects against stress and mental health issues.

DOWN AND DIRTY

There has long been a belief that playing in the dirt is good for kids. Since the late 1980s, some scientists have been saying that our obsession with sanitizing has resulted in the upswing in asthma, allergies, and eczema in children. Called "the hygiene hypothesis", the theory states that if our immune system has not had exposure to harmful bacteria, viruses, and parasites from an early age, then it unwittingly targets harmless pollen, nut proteins, or dust mites, triggering in us an allergic reaction.

This hygiene hypothesis has never been proven conclusively and while good sanitation and hand washing will always save far more lives than it will cost, it would probably be good for us to get our hands grubby with mud every once in a while.

The instant you pick up a palm of soil, your skin is coated with an invisible balm of bacteria, and research shows that this mix of soil microbes may refresh and rejuvenate our skin's natural protective shield of bacteria, helping to calm an overactive immune system. While potentially protective on unbroken skin, keep any cuts, grazes, and inflamed areas clean and covered, because sinister bugs also lurk in the dirt.

But more than this, bacteria from soil and vegetation actually suffuse the air above gardens and green spaces. Research shows that such microbes seem to give out anxiety-easing, mood-boosting vapours, and those that we swallow may help restore balance to our very own gut bacteria, keeping our immune system strong and our digestion running smoothly. This is especially important for young children, who need bacteria in the digestive tract to help stimulate the initial production of immune cells and their normal function, 70 per cent of which are located in the gut.

The closer we get to the earth and the weeds, the greater the number and the diversity of these life-giving germs that waft over and through us. Compare this to the pollutant-saturated air of our city streets and workspaces, and it's easy to see why we instinctively like to get down in amongst the dirt and the weeds – whether it be with a trowel or a tonka truck in hand.

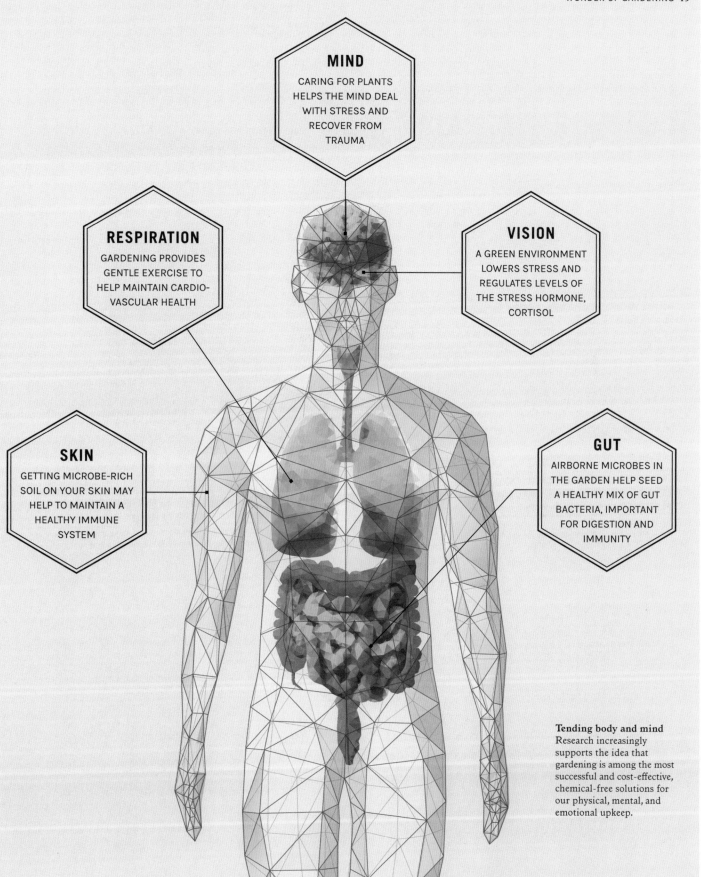

MIND
CARING FOR PLANTS HELPS THE MIND DEAL WITH STRESS AND RECOVER FROM TRAUMA

RESPIRATION
GARDENING PROVIDES GENTLE EXERCISE TO HELP MAINTAIN CARDIO-VASCULAR HEALTH

VISION
A GREEN ENVIRONMENT LOWERS STRESS AND REGULATES LEVELS OF THE STRESS HORMONE, CORTISOL

SKIN
GETTING MICROBE-RICH SOIL ON YOUR SKIN MAY HELP TO MAINTAIN A HEALTHY IMMUNE SYSTEM

GUT
AIRBORNE MICROBES IN THE GARDEN HELP SEED A HEALTHY MIX OF GUT BACTERIA, IMPORTANT FOR DIGESTION AND IMMUNITY

Tending body and mind
Research increasingly supports the idea that gardening is among the most successful and cost-effective, chemical-free solutions for our physical, mental, and emotional upkeep.

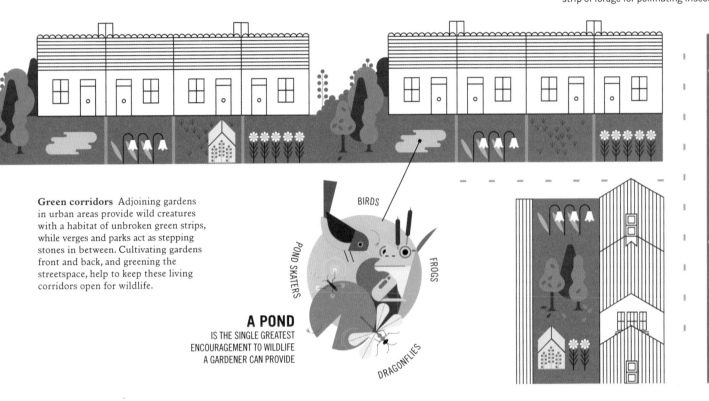

Flowers growing in central reservations provide a connecting strip of forage for pollinating insects

BIRDS

POND SKATERS

FROGS

A POND
IS THE SINGLE GREATEST
ENCOURAGEMENT TO WILDLIFE
A GARDENER CAN PROVIDE

DRAGONFLIES

Green corridors Adjoining gardens in urban areas provide wild creatures with a habitat of unbroken green strips, while verges and parks act as stepping stones in between. Cultivating gardens front and back, and greening the streetspace, help to keep these living corridors open for wildlife.

HOW IMPORTANT ARE GARDENS FOR WILDLIFE?

The unstoppable creep of concrete, metal, glass, and asphalt is pushing plant and animal life ever further to the margins. Whether it's a field or a window box, your garden can become a valuable green haven, teeming with a surprising diversity of life.

In the US alone, every single minute, more than three acres of open green space is bulldozed or paved over – and without plants there will be no wildlife. The plants in our gardens give creatures big and small food, shelter, and a place to reproduce and raise their young.

Decades of research has shown that even the plainest of gardens can be home to a galaxy of life: over 8,000 species of insect alone were detected over a thirty-year period in one modest-sized English garden. Researchers totting up the numbers of invertebrates (insects, worms, lice, millipedes, slugs, and snails) in urban and rural gardens across the world have

found a similar richness of life everywhere they have looked.

Whether it's a wide field or a window box, your garden can be a vital resource for wildlife, seen and unseen. Research shows that both small and large gardens can host a similar diversity of insect life. You decide what grows in your garden, and these choices can have a positive impact on the biodiversity that it supports.

GARDENING WITH WILDLIFE IN MIND

Choose to plant flowers or trees that blossom and pollinators will buzz in from miles around to feed on their nourishing

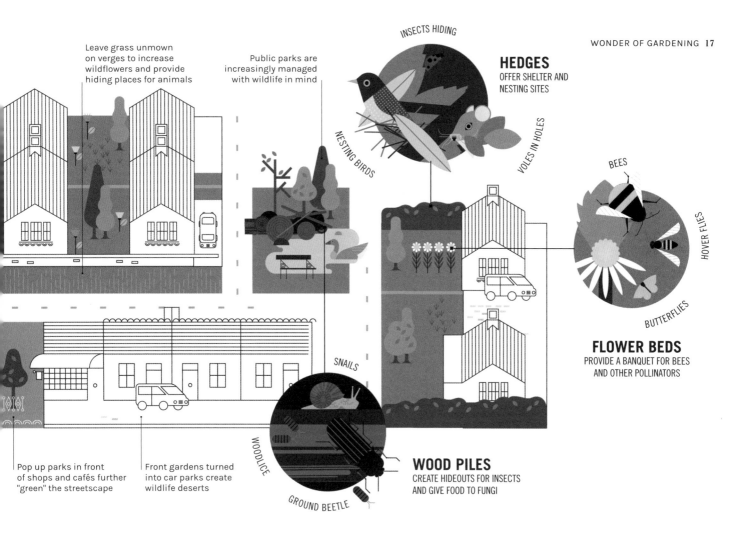

Leave grass unmown on verges to increase wildflowers and provide hiding places for animals

Public parks are increasingly managed with wildlife in mind

INSECTS HIDING

HEDGES
OFFER SHELTER AND NESTING SITES

NESTING BIRDS

VOLES IN HOLES

BEES

HOVER FLIES

BUTTERFLIES

FLOWER BEDS
PROVIDE A BANQUET FOR BEES AND OTHER POLLINATORS

SNAILS

WOODLICE

GROUND BEETLE

WOOD PILES
CREATE HIDEOUTS FOR INSECTS AND GIVE FOOD TO FUNGI

Pop up parks in front of shops and cafés further "green" the streetscape

Front gardens turned into car parks create wildlife deserts

---

nectar and pollen. Plant a hedge or grow climbers up walls and fences to provide yet more flowers, along with cover and nesting sites for birds, and shelter for mammals and insects too. Berries and rose hips will feed birds and a water bath will give them and honeybees a place to drink. A small pond or area of standing water will give a breeding site for many insects as well as frogs, toads, and newts. Leave some wood, twigs, and fallen leaves to rot down in one corner and a legion of beetles, woodlice, ants, centipedes, fungi and more will love you for it.

A patch of uncut grass will offer up a mini jungle for all kinds of wildlife, and providing you steer clear of weedkillers and insecticides, research shows that even immaculate lawns and pristine flower beds can be valuable homes for wildlife, with the rich soil a fertile land for earthworms and other soil organisms. Put simply, anything is better than concreting over your yard.

BENEFICIAL CREATURES

The other great news for gardeners is that encouraging wildlife onto your plot has many perks. The songbirds, bats, shrews, toads, and hedgehogs that we love all rely on creepy-crawlies as their main source of food, and help to keep the populations of snails, slugs, and insect plant pests under control. Some insects, like ladybirds and hoverflies, will also prey voraciously on the sap-sucking aphids that can plague new shoots in spring.

Worms and an army of other invertebrates are just some of the decomposers that recycle dead plants and animals, fallen leaves, and all of nature's waste into nutrients that sustain plant growth. Flying invertebrates – bees, flies, wasps, beetles, moths, and butterflies – pollinate over 80 per cent of all flowering plants on Earth, enabling fruit and seed production and helping to guarantee you a good crop of peaches, apples, and melons, to name but a few.

CAN GARDENING HELP SAVE THE PLANET?

As global temperatures inch higher and weather becomes more extreme, the tsunami of man-made environmental catastrophes can feel overwhelming. Everyone who tends a garden has a choice: to be part of the problem, or a powerful driver of the solution.

At no point in the last 65 million years, when an asteroid wiped out the dinosaurs, has the climate been changing as quickly as it is today. Two hundred and fifty years of burning fossil fuels, chain-sawing forests, spilling toxic chemicals, ploughing fertile soils into dust, and draining the planet of its fresh water are now coming back to bite us.

CHOOSE PLANTS THAT MAKE A DIFFERENCE

Whether a window box or a spacious meadow, the way you use your little patch on the planet really can make a difference. A small balcony can be a flower-filled oasis in a desert of concrete, while a bigger rural garden can be an equally valuable resource among endless fields filled with a single crop. Pack your plot with sources of sugar-rich nectar year-round and you will help to feed populations of pollinating insects (see pp.16–17), struggling with the loss of their natural habitat and widespread use of pesticides.

Plants will thrive and need less attention when they are planted in the right spot (see pp.58–59). For example, growing plants that are adapted to arid conditions in the driest part of your garden will reduce the need for watering and so save on this increasingly precious commodity.

Where you buy plants can also have an environmental impact, because you can choose plants propagated at a local nursery, where you can ask the staff about their use of synthetic fertilizers and pesticides, or you can opt for

HIDDEN COST
CO2 EMISSIONS FROM PRODUCING A SQUARE METRE OF CONCRETE ARE THE SAME AS BURNING 100 LITRES OF PETROL

PAVED OVER
PRIORITIZING PARKING OVER PLANTS REDUCES THE ABILITY OF GARDENS TO ABSORB CO_2, POLLUTANTS, AND RAINFALL

Solid paving causes run-off after rain, which can result in flooding

Bare walls and fences don't reduce air pollution or provide for wildlife

GREEN UP
ADDING PLANTS TO ANY OUTDOOR SPACE BRINGS BOTH ENVIRONMENTAL AND HEALTH BENEFITS

Trees absorb CO_2 and pollution from passing traffic

Green roofs can be planted on car ports or sheds

Permeable paving reduces run-off after heavy rainfall

Flowering plants provide nectar for insects and boost your mood

Windowboxes and climbers add planting to every available space

Hedges reduce air pollution and are valuable for wildlife

mass-produced, imported plants grown in heated greenhouses and transported hundreds of miles to the garden centre or DIY store shelves. Raising your own plants from seed minimizes the cost both to you and the planet, giving you control over the resources used while allowing you to enjoy the entire life of a plant.

VALUE YOUR OWN GREEN SPACE

Garden design can matter too. Think carefully before swapping real turf for plastic or concreting over front gardens to create driveways. For a start, concrete has a vast environmental cost: making a square metre can emit up to 260kg (575lb) of CO_2, equivalent to burning over 100 litres (22 gallons) of petrol. Such a choice also increases the risk of flooding for you and your neighbours (see p.23), and squanders the valuable potential of the plants and other life in your plot to capture carbon dioxide along with other harmful pollutants from the air (see p.22 and p.21), and to cool your surroundings in hot summer weather (see p.22).

Life has an incredible ability to find a way, but there is a great deal you can do to stack the odds in its favour. Looking after your soil and growing plants that are welcoming to wildlife will mean that insects, birds, and many other creatures will make their way to your little sanctuary (see pp.16–17). The positive impact of your garden and gardening will ripple out far further than you might ever have thought possible.

DO PLANTS ABSORB POLLUTANTS FROM THE AIR?

Nine out of ten of us live in air so polluted it is harming our health. Air pollution kills between seven and nine million people worldwide every year, and causes heart and lung disease, stroke, diabetes, cancer, dementia, as well as headaches and anxiety.

Poisonous nasties floating in our air include the lung-damaging greenhouse gas nitrous oxide, toxic volatile organic compounds (VOCs), and lethal invisible dust, known as particulate matter. Plants can't rid our streets and homes of these toxic fumes, but they can draw a sizeable proportion of the pollution from exhausts and chimneys out of harm's way, meaning that a garden can have a particularly powerful positive impact if you live in a town or city.

NATURAL AIR FILTERS

Plants neutralize the harmful effects of some dangerous gases, such as nitrous and sulphur oxides, when they are taken in through tiny pores on the undersides of leaves (stomata), and dissolve into the water within the leaf. Microscopic hair-like bristles (called trichomes) that coat plant leaves and stems effectively filter the air by trapping fine particulate matter, which is then washed away by rain. Species with small, lobed leaves slow passing air and filter

pollution most effectively; conifers like Chinese juniper (*Juniperus chinensis*) with lots of small, needle-like leaves are best for purifying air. Nevertheless, all plants, even grass, will capture some air pollution. The more you grow – especially between you and roads – the cleaner the air will be.

WHAT'S THEIR IMPACT INDOORS?

Since NASA research in the 1980s found evidence that plants can purify the air indoors, people have raved over their ability to remove the toxic chemicals in cleaning products, air fresheners, fabrics, and carpets from our homes. While it's true that both leaves and microbes within plants absorb this pollution, these claims now seem to have been overblown: in reality only an indoor jungle would remove enough VOCs to make a real difference. But plants do bring many health benefits, including lifting mood, and improving concentration and overall wellbeing (see pp.14–15).

Pollution traps The leaves of different plants have a variety of features that allow them to catch and hold pollutants from the passing air.

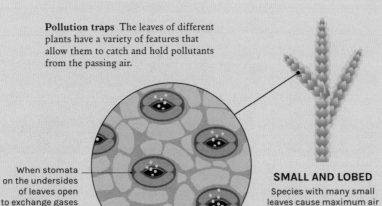

When stomata on the undersides of leaves open to exchange gases they also take in and trap pollutants

SMALL AND LOBED
Species with many small leaves cause maximum air turbulence and have the largest pollution-capturing surface area.

WAXY
Many leaves have a waxy surface that is sticky enough to trap and retain fine particulate matter from the air.

HAIRY
Fine leaf hairs slow down airflow and capture the toxic particules contained in smoke and vehicle exhaust fumes.

CAN MY GARDEN CAPTURE AND STORE CARBON DIOXIDE?

It may not be possible to plant the estimated 700–1,000 trees annually that might be needed to offset all your carbon emissions, but you can plant and manage your plot in ways that maximize its ability to store and retain carbon dioxide.

Carbon dioxide is chief among the greenhouse gases that are turning our planet into a sauna. Every living thing breathes it out and it is also released whenever plant matter, including fossil fuels, is burned.

THE ROLES OF PLANTS AND SOIL

Unique among living things, plants remove carbon dioxide from the air to make their own food during photosynthesis (see pp.70–71), turning airborne carbon into sugars and starches, and locking it away in carbon-rich tissues, like wood. When plants die and decay they become food for decomposing organisms, including soil-dwelling bacteria and fungi (see pp.44–45), which break down plant matter as they feed on it. This releases some carbon back into the atmosphere, as part of a process called the carbon cycle, while the rest is incorporated into the soil,

and the bodies of the countless organisms living within it. Soils that are fed annually with organic matter and where digging is minimized (see pp.42–43) will store more carbon than those that are regularly tilled.

Every species of garden plant will hoover up carbon dioxide at a different rate, and this varies with temperature, humidity, and rainfall. Fast-growing shrubs and trees are especially good at extracting carbon dioxide from the air, quickly turning it into wood, where it will be locked away for decades. Saplings are slow to capture carbon and only get into their stride when five to ten years old. A mature tree might draw down 20–35kg (44–79lb) of carbon dioxide each year (more in tropical regions) – the same as the amount emitted by burning just 10–15 litres (2–3 gallons) of petrol.

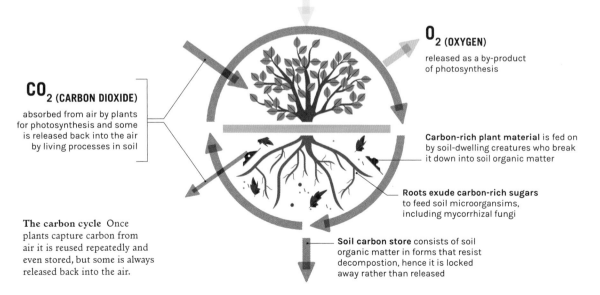

SUNLIGHT
powers photosynthesis in plants

O_2 **(OXYGEN)**
released as a by-product of photosynthesis

CO_2 **(CARBON DIOXIDE)**
absorbed from air by plants for photosynthesis and some is released back into the air by living processes in soil

Carbon-rich plant material is fed on by soil-dwelling creatures who break it down into soil organic matter

Roots exude carbon-rich sugars to feed soil microorgansims, including mycorrhizal fungi

The carbon cycle Once plants capture carbon from air it is reused repeatedly and even stored, but some is always released back into the air.

Soil carbon store consists of soil organic matter in forms that resist decompostion, hence it is locked away rather than released

CAN PLANTS KEEP YOU COOL?

Wherever trees have been traded for skyscrapers and grasslands for paving
and roads, temperatures will invariably be higher. Plants can provide relief
by offering shade and a breath of cool, refreshing air.

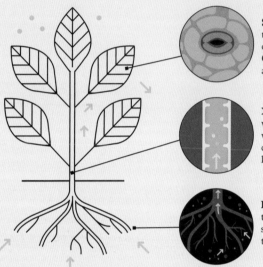

Transpiration Water
absorbed by roots moves
up through xylem vessels
in the stem to leaves, where
it evaporates through open
stomata, cooling the air
around the plant.

Stomata are pores on
the lower leaf surface that
open in daylight, allowing
CO_2 in and water vapour
and O_2 out.

Xylem are the channels
within plants that conduct
water upwards by simple
capillary action as it is
lost from leaves.

Roots draw water into
their tissue from the
surrounding soil to supply
the transpiration process.

Summer in the city isn't always a nice place to be.
Heat builds in expanses of concrete and asphalt to
create oppressively hot days and stifling nights. This
"urban heat island" sees us dashing for the comfort
of air conditioned buildings. Adding plants to the
situation can reverse some of these effects.

NATURAL CLIMATE CONTROL

A canopy of leaves that creates shade from the
midday sun in summer, could make you feel up to
10–15°C (18–27°F) cooler, while trees to the south
and west of a house shield it from the midday and
afternoon sun, with a cooling effect that could slash
air conditioning costs by as much as 50 per cent.

But plants offer more than a green parasol. As
water evaporates, it lowers the temperature nearby,
which is why people flock to fountains and open
water in summer. Plants cool the air in a similar
way by drawing water up from soil into their leaves,
where it gradually evaporates through little pores

(stomata) on their undersides. This process, called
transpiration, cools both the plant and surrounding
air in much the same way as sweat cools our skin.

GREENING TURNS DOWN THE HEAT

A planted rooftop (aka "green roof") can have
profound cooling effects inside a building. Dark-
coloured roofs absorb the sun's energy, but carpeting
them with plants exudes cooling moisture and
provides a shield from the sun's baking rays. This
reduces summer highs of 57°C (134°F) to 30°C
(86°F), and buffers up to 95 per cent of heat that
would permeate to upper floors, reducing the energy
needed for indoor cooling by up to 25 per cent.

Indoor plants can also have a cooling effect, and
the more of them the better. Green walls festooned
with plants are increasingly popular in public and
commercial buildings as an environmentally friendly
way to improve air quality and lower temperatures.
Call it green living in every sense!

CAN GARDENS AVERT FLOODING?

With increasingly erratic weather, flooding is likely to become a more common event in many areas. Gardens, or a lack of them, play a surprisingly big part in whether extreme rainfall leads to flooding.

Between 1980 and 2016, the number of storms in Europe doubled and the rate of flooding more than quadrupled. Floods happen when rain falls faster than it can drain away. Paved surfaces can't absorb water, forcing it to flow into drains and gullies, potentially overloading them. Garden soil, on the other hand, allows water to percolate downwards, taking the pressure off drainage systems during storms. Any garden helps to prevent runoff and mitigate the risk of floods, but how well it does this is influenced by the amount of hard landscaping and the type and quality of its soil (see pp.38–39).

FLOODING: IT'S ALL IN THE SOIL

Sandy soils have large particles with generous gaps (pores) between them, and so water can pass through easily. Clay soils are the opposite, each particle swelling and holding onto water like a sponge, slowing the drainage of rainwater.

Soil that has been squashed down or compacted, typically by vehicles, people, or animals, is less porous and more liable to flood. In compacted soil, there are no channels between particles for water to drain away. Forking the surface (see pp.126–127) makes only temporary drainage channels and the best way to prevent water pooling and running off is to nurture a healthy soil structure by minimising digging and by mulching regularly with organic matter (see pp.42–43). Canopies of leaves also slow the rate that water strikes the ground, and roots grip the soil, preventing it from washing away (eroding).

A GARDEN TO CAPTURE RAIN

Rain gardens are attractive planted areas designed to slow down and collect water running off paved areas, and even roofs, during heavy rainfall. Water collects in a basin that has been dug in a well-drained area, to prevent water pooling elsewhere in the garden and nearby drains becoming inundated. They are a simple and effective alternative to more conventional ditch or pipe drainage systems, and are usually cheaper and easier to install.

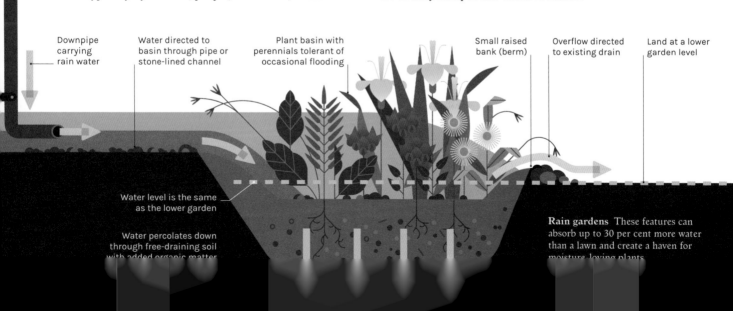

Downpipe carrying rain water

Water directed to basin through pipe or stone-lined channel

Plant basin with perennials tolerant of occasional flooding

Small raised bank (berm)

Overflow directed to existing drain

Land at a lower garden level

Water level is the same as the lower garden

Water percolates down through free-draining soil with added organic matter

Rain gardens These features can absorb up to 30 per cent more water than a lawn and create a haven for moisture-loving plants.

HOW WILL CLIMATE CHANGE AFFECT MY GARDEN?

The planet is warming and gardeners are on the front line, witnessing first-hand the effects of the resulting shifts in weather patterns on what they grow. Rather than despair, seize the chance to learn more about plants and hone your growing skills.

Apple blossom Earlier blossom is vulnerable to damage by spring frosts and may bloom before insect pollinators are active, reducing yields of fruit.

OUT EARLY

MILDER WINTERS AND WARMER SPRINGS STIMULATE FRUIT TREES TO FLOWER EARLIER IN THE YEAR

Temperatures worldwide have risen by a little over 1°C (1.8°F) since humans started burning fossil fuels some 170 years ago. This might sound like the merest of blips on a thermometer, but the consequences are profound, upsetting the fine balance of global weather systems and making extreme weather, like droughts, more common.

SHIFTING THE SEASONS: A RISKY CHANGE

As the planet heats up, spring is arriving earlier in many parts of the globe, causing plants and animals to emerge from their winter hibernation sooner. In the Arctic, spring has advanced 16 days in a decade; in North America, blueberries in Massachusetts flower three to four weeks earlier than in the mid-

1800s, while in the UK, spring is already arriving a whole month earlier than it did 30 years ago.

Warmer temperatures make light sensors (known as photoreceptors) dotted throughout plant tissues more sensitive to sunlight, triggering them to end dormancy earlier, when days are still short. Nothing can be counted on with climate change though – in some areas of the US (Florida and Texas) spring has actually started arriving later. These unpredictable shifts will put plants in some regions at greater risk of being caught out by a late frost (see graphs opposite), which can kill tender plants already in the ground and damage the soft new shoots and early blooms of hardy plants.

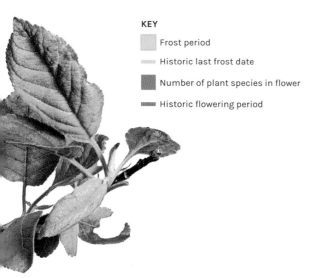

KEY

- ▢ Frost period
- ▤ Historic last frost date
- ▢ Number of plant species in flower
- ▬ Historic flowering period

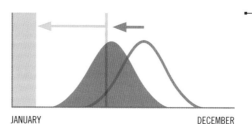

JANUARY   DECEMBER

PRE-1960

A stable climate meant little overlap between flowering and frost periods, resulting in limited frost damage.

JANUARY   DECEMBER

UK – TODAY

Increased risk of frost damage because flowering dates have advanced faster than the frost period has receded.

Changing risk of frost damage

With a warming climate, plants tend to flower earlier in the year, but last frost dates have not changed in sync with flowering. These factors have altered the risk of frost damage unevenly in different countries.

JANUARY   DECEMBER

US – TODAY

Reduced risk of frost damage because the frost period has receded faster than flowering dates have advanced.

ADAPT FOR SUCCESS

Much of the natural world will adapt to the changes that lie ahead and gardeners should try to emulate nature's versatility. Some traditional planting advice may need to be cast aside because a changing climate will bring new challenges and opportunities. Longer growing seasons offer a chance to try more adventurous plant selections: many gardeners in northern climes might be excited at the prospect of growing plants previously more at home in Mediterranean regions or lower latitudes.

For gardeners lucky enough to live in temperate areas, a longer growing season may also help grow food crops, giving vegetables and fruit more time to swell, ripen, and yield bigger harvests. Unfortunately, alongside these exciting planting possibilities will come waves of new pests and diseases. Warmer, wetter conditions favour fungi, bacteria, and insect pests, allowing new species to spread northwards into regions once too cold for them to thrive (see pp.194–201).

Even top climate scientists aren't sure how weather systems will evolve, so little is certain other than that growing conditions will be increasingly unpredictable. Once-a-century heatwaves and flash floods will become regular fixtures. There could be years when all but the toughest of plants succumb to drought, or flooding causes waterlogged conditions.

Gardeners need to anticipate and prepare for these challenges, perhaps by collecting rainwater, creating rain gardens (see p.23), and possibly by swapping old favourites for plants that are well adapted to tolerate drought (see p.117) and other extreme weather conditions.

LAST FROST DATES
(change since 1960s)

SOUTHEASTERN AUSTRALIA

ON AVERAGE 4 WEEKS LATER

IRAN

UP TO 23 DAYS EARLIER

POLAND

UP TO 21 DAYS EARLIER

PREPARING
THE GROUND

WHAT EXACTLY IS GARDENING?

For millennia, humans have sown seeds and nurtured plants not only for
food and medicine, but also for pleasure. From the ancient gardens of Babylon,
to the perfect lawns and roses of 1950s suburbia, and today's wildlife havens,
we have gardened to express ourselves.

In a garden, everyone has the freedom be a designer, artist, technician, and scientific observer – your own space in the world to plant, prune, and use as you please. Like a zookeeper, you can delight in selecting plants from across the globe and giving them a place to call home.

MOULDING NATURE

Gardening is about delighting in the best that the plant world has to offer. However, creating a perfect green escape of voluptuous blooms and bountiful harvests usually means meddling with the natural order of things. Throughout history gardening has meant taming nature to create a world quite unlike the one in which plants evolved. Outside gardens, the wild plants you see create a constantly changing tapestry, woven by evolution and the daily battle for survival. Each is jostling for its place, trying to win out over its neighbours. Wherever you live, the natural rhythm of life is the same: a bare patch of earth is rapidly covered by "pioneer" species whose seeds float on the breeze, such as fast-growing grasses and annuals like groundsel. The death and decomposition of these first colonisers enriches the soil, setting the stage for larger plants to move in and find their niche. Left undisturbed, this natural "succession" of plants often eventually produces woodlands or forest.

Naturalistic planting styles
Combinations of airy grasses and vibrant flowers mimic the appearance of natural plant ecosystems, but usually contain plants that would not be found together in the wild.

PRAIRIE
A CAREFULLY MANAGED MIX OF GRASSES AND LATE-FLOWERING PERENNIALS INSPIRED BY TEMPERATE GRASSLANDS

MEADOW
CUT ANNUALLY, AFTER FLOWERS AND GRASSES HAVE DROPPED SEEDS, THIS IS A MORE SELF-SUSTAINING DISPLAY

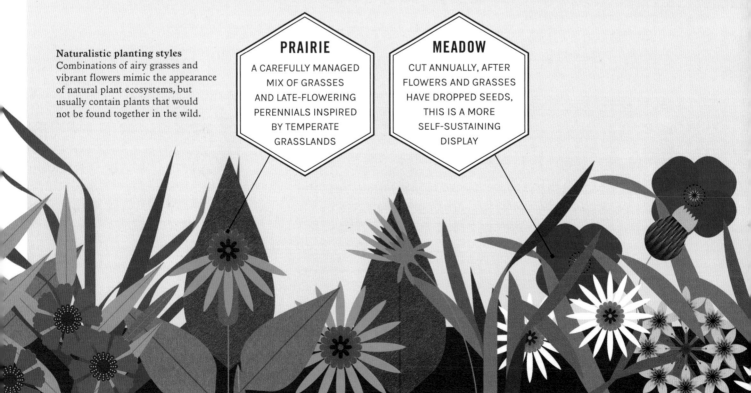

THE EVOLUTION OF GARDENS

In our gardens, each plant is put in its place like a piece of furniture, and plant succession is stamped on. Pioneer species are called weeds and are plucked out or poisoned. Gardening fashions and ideas have never stood still, but no matter the style or scale, they always involve bending nature to our will to some extent. A gardener's greatest skill is maintaining their created world, helping each plant to live happily and harmoniously.

During much of the 20th century, gardeners strived for emerald-green manicured lawns and artificially lush plants pumped up with fertilizers, while keeping pests at bay with noxious sprays. By the 1980s, however, the environmental cost of this approach was becoming clear (see pp.30–31), and organic methods and plantings with a more "natural" look started to gain a following. Today, there is a strong trend towards naturalistic and even more self-sustaining planting styles, allowing perennials, shrubs, and trees to grow, spread, thrive, or fail, changing a garden's character as plant succession is restored. All of which is great news for garden wildlife and makes this a fascinating time to garden.

Forest garden This garden style echoes the plant layers – from tree canopy to ground cover – found at forest margins. It's an example of a more naturalistic, self-sustaining way of gardening.

TREES
GROWN TO PROVIDE PRIVACY, CAST SHADE, OR BEAR FRUIT, WITH BEAUTIFUL FOLIAGE, BLOSSOM, AND BARK

SHRUBS
FORM A LOWER STOREY OF FOLIAGE AND STRUCTURE. OFTEN SMOTHERED IN FLOWERS AND FRUITS

PERENNIALS
AN INCREDIBLE DIVERSITY OF VIBRANT BLOOMS AND FOLIAGE FORMS FOR ALL SEASONS AND LOCATIONS

GROUND COVER
LOW-GROWING, OFTEN SHADE-LOVING PLANTS, THAT SPREAD OVER SOIL, PREVENTING WEED GROWTH

HOW CAN I GARDEN MORE SUSTAINABLY?

Sustainability is a watchword in this era of environmental awareness that means leaving the world in a healthy state for generations to come. Not taking more than can naturally be replaced, including when we garden, is our route to a sustainable future.

For most of humanity's history we have lived sustainably, being careful to steward our land and resources so our children will be able to live life as we did. Today we take far more from our world than we put back, but by simply re-evaluating our methods we could all make our gardening much more sustainable.

REDUCE SOURCES OF POLLUTION

If you use a petrol-powered mower, leaf blower, or hedge trimmer, think seriously about whether you really need it. Unlike petrol-powered vehicles, the emissions from garden machinery are largely unregulated and their inefficient engines invariably pour out a cocktail of toxic gases. According to one Swedish study, using a mower for thirty minutes produces the same air pollution as a 50-mile car trip. If only for the health of your family's lungs, try using hand tools where practical, or switch to electric alternatives, especially if you can opt for renewable energy sources from your supplier.

Think carefully about whether you really need fertilizer (see pp.122–125). While a pack of lawn or plant feed might look harmless, synthetic fertilizers are made using energy-intensive processes and their use can pollute waterways. Soil that is regularly enriched with mulches of organic matter can supply plants with all the nutrients they need (see pp.42–43).

Pulling nitrogen gas from air and compressing it into liquid fertilizer is an epic undertaking: up to five litres (one gallon) of carbon dioxide are produced in the manufacture of just a one-gram ($^1/_{25}$ oz) droplet of nitrogen. The damage doesn't end there, because bacteria in soil turn some of this nitrogen into toxic nitrous oxide, which is an even more powerful greenhouse gas than methane or carbon dioxide.

Nutrients in fertilizer can also be washed from the soil by rain, ultimately seeping into springs, rivers, lakes, and the sea, where unnaturally high levels of nitrogen and phosphorus cause algal blooms, which in turn can lead to oxygen-starved "dead zones" where marine life is suffocated.

Chemicals available to gardeners to control weeds, insects, other animal pests, and fungal diseases also have potential to harm the wider world (see pp.52–53) and can easily have unintended consequences as they are, by definition, poisonous to life. There are often less harmful ways to limit any damage caused by pests and diseases, and gardeners are increasingly using an approach called "integrated pest management" (see p.53) to help prevent and control problems.

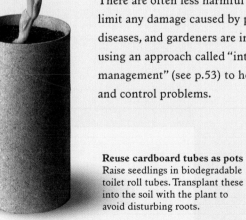

Reuse cardboard tubes as pots
Raise seedlings in biodegradable toilet roll tubes. Transplant these into the soil with the plant to avoid disturbing roots.

Sustainable choices

Make positive changes to reduce the environmental impact of gardening, while safeguarding wildlife and your own health.

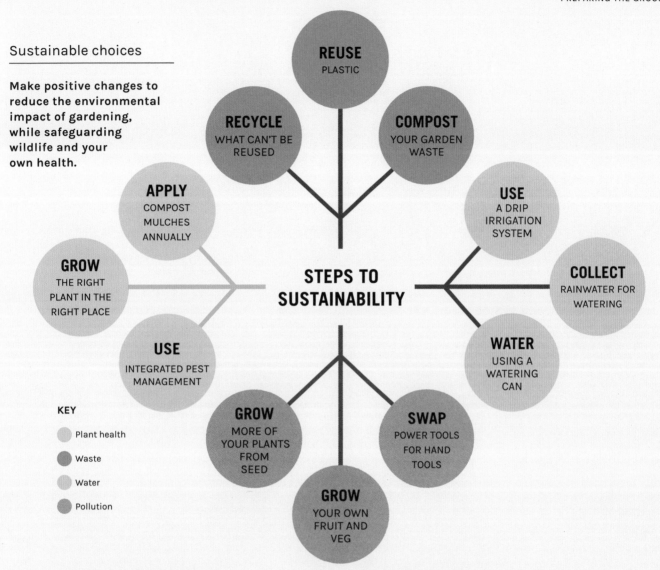

REUSE
PLASTIC

RECYCLE
WHAT CAN'T BE REUSED

COMPOST
YOUR GARDEN WASTE

APPLY
COMPOST MULCHES ANNUALLY

USE
A DRIP IRRIGATION SYSTEM

GROW
THE RIGHT PLANT IN THE RIGHT PLACE

STEPS TO SUSTAINABILITY

COLLECT
RAINWATER FOR WATERING

USE
INTEGRATED PEST MANAGEMENT

WATER
USING A WATERING CAN

GROW
MORE OF YOUR PLANTS FROM SEED

SWAP
POWER TOOLS FOR HAND TOOLS

GROW
YOUR OWN FRUIT AND VEG

KEY

- Plant health
- Waste
- Water
- Pollution

WATER CONSERVATION

Fresh water will become an increasingly precious commodity as the planet warms, and finding ways to reduce water use also tends to reduce work for gardeners. Choose plants well suited to your climate and soil type that will thrive without additional watering. A lot of energy is invested in providing clean tap water, so make use waste water from sinks and baths where possible and use butts to capture rainwater from roofs. Sprinkler systems ought to go the way of the steam engine, given they typically hose out 4,500 litres (1,000 gallons) of tap water every hour. Reduce waste by watering with a can or installing a drip irrigation system (see pp.114–115).

CIRCULAR LIVING

Reducing what we consume, reusing or repurposing items, and recycling what is left are the tenets of a "circular economy" and key to sustainability. Buy only what you need, avoid single-use plastic, and reuse all you can. Rather than buying pot-grown plants, grow plants from seed in old plastic pots or containers created from recycled food tubs, paper, or cardboard tubes. Growing your own fruit and veg can cut food bills and reduce plastic packaging, food miles, and food waste. Gardeners value homegrown food and won't let it go to waste. Compost what isn't eaten or given away, along with garden waste, to give mulch to feed soil for future crops (see pp.188–191).

WHICH WAY DOES MY GARDEN FACE?

Gardeners like to talk about the direction their garden faces or its "aspect". A patch of land doesn't really have a face, of course, but it is useful to know whether your yard gets lots of sunlight or is shaded by your house for much of the day.

As the sun swings through the sky, the shadows cast by buildings and tall trees tilt, shrink, stretch, and turn so that some parts of your garden get much more sun than others. Unless you live on or near the equator (between the tropics of Capricorn and Cancer), the midday sun will never be directly overhead, but will shine down on you at an angle – if you are in the northern hemisphere, the sun will always be in the southern sky, even if only slightly in summer, and because of this, your house's shadow will fall northwards.

ASSESSING ASPECT

The way your garden faces is simply the compass direction in which it extends away from your house: with your back against your house and compass in hand, look out at your garden to see which way the arrow points. If you have garden space both to the front and rear of your house, they will face in opposite directions and may be quite different in character as a result.

If you are unlucky enough to have a garden that faces north, then direct sunlight will be in short

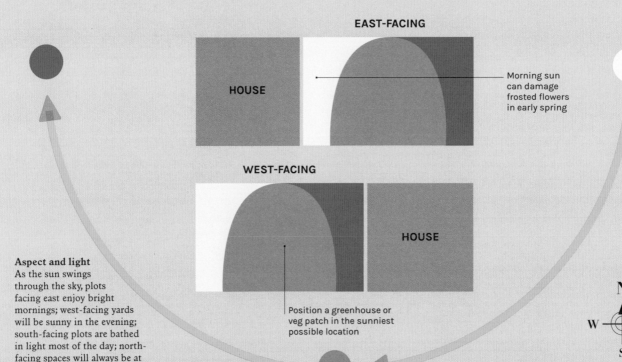

EAST-FACING

HOUSE

Morning sun can damage frosted flowers in early spring

WEST-FACING

HOUSE

Position a greenhouse or veg patch in the sunniest possible location

Aspect and light
As the sun swings through the sky, plots facing east enjoy bright mornings; west-facing yards will be sunny in the evening; south-facing plots are bathed in light most of the day; north-facing spaces will always be at least partly shaded by the house.

N
W E
S

supply. Where the compass points south, a garden is south-facing, which means it will receive sun for much of the day, making you the envy of fellow gardeners. East-facing gardens get the best of the morning sun (and are shaded in the evening), while west-facing gardens enjoy the warm evening sunset.

CONSIDER EVERY DETAIL

Fences or walls on a garden's boundary also have their own aspect and cast their own shade. A northern boundary "faces" the south and will be sunny throughout the day. These south-facing sun traps make prime planting spots, because fences and walls both absorb and reflect the sun's energy, providing extra warmth that suits heat-loving plants and helps ripen fruit on wall-trained trees.

Of course, gardens rarely line up exactly with the compass points and the amount of shade will also be influenced by nearby buildings, trees, and high hedges. The size of your plot is also significant,

because small gardens tend to have less open space around them and are often shaded by nearby trees and structures. To find out the sunny and shady parts of your garden, it is a good idea to take photos at different times of the day.

Let aspect influence design

KNOWING A GARDEN'S ORIENTATION HELPS TO PLAN THE BEST LOCATION FOR EACH PLANT AND KEY DESIGN FEATURES

The best way to maximize your garden's potential is to understand its aspect, because you can only position a seating area to be bathed in evening sunlight or shaded from midday heat if you know where the sun and shade fall at different times of day. Plants are also adapted to thrive in in a range of conditions, so plant those that need full sun and warmth in south-facing positions and reserve north-facing spots for shade-loving plants.

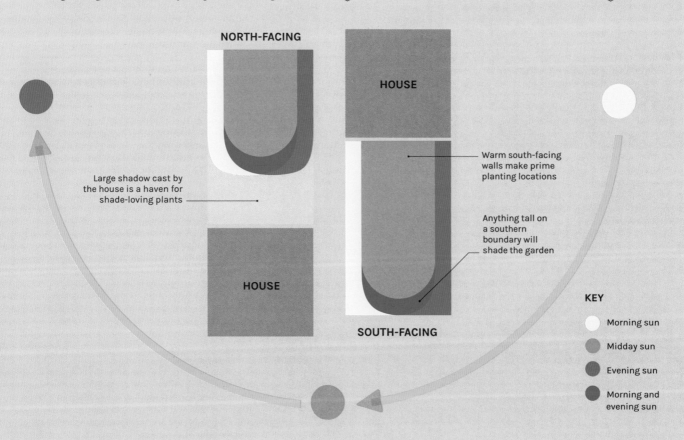

NORTH-FACING

Large shadow cast by the house is a haven for shade-loving plants

HOUSE

HOUSE

Warm south-facing walls make prime planting locations

Anything tall on a southern boundary will shade the garden

SOUTH-FACING

KEY

○ Morning sun

● Midday sun

● Evening sun

● Morning and evening sun

HOW DOES THE WEATHER AFFECT WHAT I CAN GROW?

Although the needs of individual plants vary, sunlight, rainfall, and heat are all crucial to the reactions that power plant growth. With this in mind, it's easy to see why the weather in your area – your local climate – influences what it's possible to grow.

Regardless of how well you garden, the weather is something over which you have no control. It affects the conditions in your plot and the plants you'll be able to grow successfully outdoors.

LIGHT BRINGS LIFE

The amount of sunlight your garden gets over the year sets a limit on which plants you can grow. Sunlight allows plants to make food, and the more that falls on a plant's leaves, the more fuel (sugar) they will have for growth. How long the sun stays above the horizon is largely dictated by latitude – how far you are from the equator. The further from the equator, the shorter the growing season: in

Iceland, for example, fewer than 1,400 hours of sunshine will light up your plot each year, which is less than half the hours that a garden in Greece would get.

Some regions experience much more cloud cover than others, which also influences the amount of sunlight plants receive. Garden plants that flourish in bright sunlight, such as rock rose (*Cistus*), usually originate from hot, sun-baked climates. The leaves of many others, such as Japanese maple (*Acer palmatum*), are adapted to dull conditions and ill-equipped to cope with blazing sun. Day length is also used by some plants as a signal for flowering (see pp.142–143).

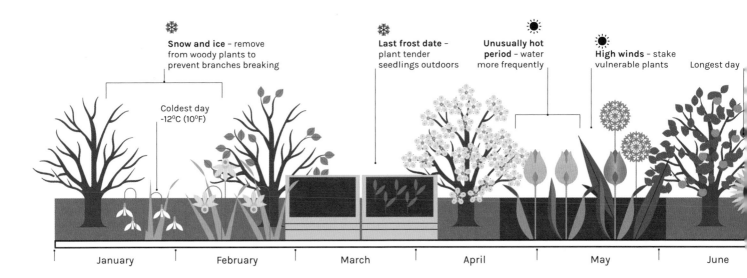

Snow and ice – remove from woody plants to prevent branches breaking

Coldest day -12°C (10°F)

Last frost date – plant tender seedlings outdoors

Unusually hot period – water more frequently

High winds – stake vulnerable plants

Longest day

January | February | March | April | May | June

A year in your garden Observing, measuring, and recording the weather can help decide what to plant and when it's best to sow, plant, and harvest.

WATCH THE TEMPERATURE

If sunlight gives fuel to power the machine, it's temperature that affects the speed at which the cogs spin. Almost all of life's chemical reactions are controlled by special molecules called enzymes, which are calibrated to specific temperatures for each plant species. Enzymes usually work faster at warmer temperatures, but their limits vary between plant species. In broccoli, for example, enzymes are calibrated such that growth almost ceases above 25°C (77°F). But in all plants, the green chloroplasts, where photosynthesis happens (see pp.70–71), stop working when temperatures hit 40°C (104°F), meaning death is not far away unless conditions cool. Fruits ripen faster and taste sweeter in warm conditions, which speed up the activity of ripening enzymes.

Low temperatures quickly kill plants that are not evolved to cope with freezing, and gardeners rightly pay close attention to minimum winter temperatures (see pp.80–81). Even if your region rarely experiences freezing temperatures, plants whose biological machinery is tuned to warm temperatures may struggle. Citrus trees, for example, are unable to flower where night-time temperatures drop below 10°C (50°F) in winter.

WIND AND RAIN

Wind can cause damage by breaking stems and branches, but gardeners are often unaware of its harmful drying effects. Moisture continually evaporates into the air from leaves (transpiration), and high winds speed up this evaporation, quickly dehydrating leaves, as if under a hand drier. Needle-like leaves have a smaller surface area and so suffer this effect much less.

Rainfall gives plants essential water, but each will have different needs depending on where they evolved. Average rainfall data gives a useful guide for your area, but seasonal variations also affect what will thrive in your garden. Some plants are extremely tolerant of wet soil and need to remain moist, while others can only survive high rainfall in extremely fast-draining soil, or will struggle during wet winters. Plants in wet climates are also more prone to fungal diseases, and damage caused by slugs and snails.

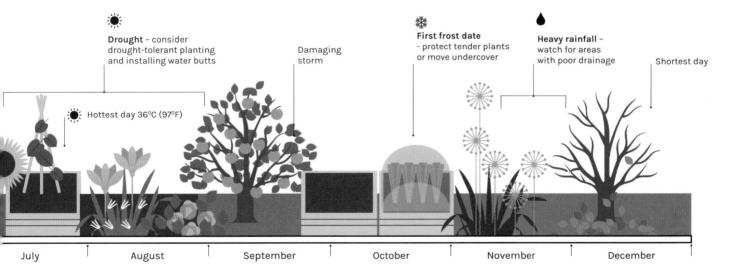

Drought – consider drought-tolerant planting and installing water butts

Damaging storm

First frost date – protect tender plants or move undercover

Heavy rainfall – watch for areas with poor drainage

Shortest day

Hottest day 36°C (97°F)

July   August   September   October   November   December

WHAT IS A MICROCLIMATE?

Every garden has a unique combination of growing conditions; its own quirks that affect where sunlight and rain fall, how wind buffets off nearby buildings and trees, and where cold air gathers. Each of these is called a microclimate.

It can take years to fully unravel the idiosyncrasies of your garden's microclimates, yet getting a handle on them will enable you to select the right plant for the right place (see pp.58–59) and find the best spot to lounge on a sunny afternoon!

LOOK TO THE SUN

The sun is the ultimate giver of life and the direction in which your garden "faces" – its aspect (see pp.32–33) – will have a bearing on everything else. Where light falls as the sun first peeks above the horizon and sets in the evening will be unique to your garden, and the way the shadows shift as it arcs across the sky may surprise you: an area in direct sunlight in summer may be in shade during winter.

Buildings, walls, and fences also affect microclimates. Sheltered areas near south-facing walls are termed "sun traps", and can be an entire hardiness zone (see pp.80–81) higher than the rest

of the garden. A south-facing wall in a mid-latitude, northern hemisphere garden will gain up to 3.5 times more light and heat from the sun than the wall's north side, making it ideal for tender plants or those with fruit to ripen. Walls and fences also shelter soil at their base from rainfall, creating a dry "rain shadow" at their base.

EFFECTIVE SHELTER

Wind is often one of the most significant stresses on plants, especially in gardens facing prevailing winds, near the coast, or on exposed hilltops. The turbulence produced by trees, walls, and buildings is often so unpredictable that even supercomputers can struggle to forecast how wind will swirl. Windbreaks in the form of trees, hedges, fences, and walls can filter or block the force of the wind to protect plants.

Existing vegetation, especially trees, creates its own unique microclimates, casting shade, increasing

Where to find microclimates

Watch how the features in your garden interact with the forces of nature to produce microclimates that will influence plant growth.

RAIN SHADOW
Rain usually falls at an angle, blown by the wind. Where a wall or fence blocks its path, soil on the leeward side stays dry.

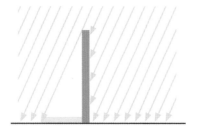

HEAT STORE
When sunlight strikes a wall, some of its energy is absorbed. This stored heat slowly radiates out at night, after a sunny day, providing a cosy spot for plants.

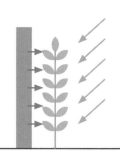

nearby air humidity (see p.22), and draining valuable moisture from surrounding soil. Conditions beneath deciduous trees change with the seasons. Where the leaf canopy creates cool shade in summer could be in full sun once leaves fall in autumn.

FROST POCKETS

Pay attention to your local landscape (topography) to help you better understand how prevailing winds might be channelled into your garden and where cold air might settle. Dips, valley bottoms, and solid walls or fences at the bottom of slopes are where cool air collects at night causing particularly cold areas, termed "frost pockets".

SHELTER FROM WIND

The best windbreaks filter and slow airflow rather than block it. Hedges and slatted fences filter out 50–60 per cent of the wind, while solid walls and fences cause damaging eddies.

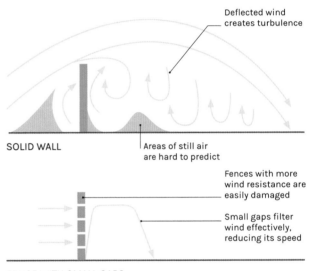

Deflected wind creates turbulence

SOLID WALL

Areas of still air are hard to predict

Fences with more wind resistance are easily damaged

Small gaps filter wind effectively, reducing its speed

FENCE WITH SMALL GAPS

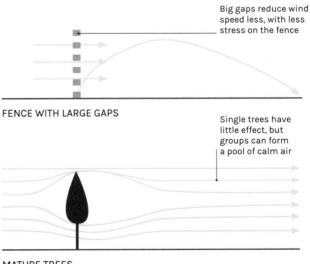

Big gaps reduce wind speed less, with less stress on the fence

FENCE WITH LARGE GAPS

Single trees have little effect, but groups can form a pool of calm air

MATURE TREES

LOW-LYING FROST

Cold air is denser than warm air, causing it to move down slopes and gather in dips, or where it meets a barrier. Frost is slow to lift from these cold pockets, damaging all but the hardiest plants.

VALLEY BOTTOM

AGAINST A HEDGE

WHAT IS SOIL?

You probably don't think much about the dirt beneath your feet, but the muck that gets under your fingernails is actually nature's secret sauce; a precious resource for plants and wildlife that's so much more than the sum of its parts.

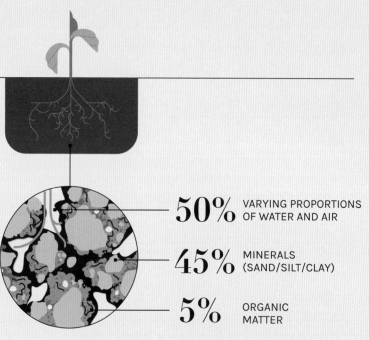

50% VARYING PROPORTIONS OF WATER AND AIR

45% MINERALS (SAND/SILT/CLAY)

5% ORGANIC MATTER

For centuries scientists have distilled soil down into its most basic chemicals and components: a blend of rocks, minerals, decayed plant and animal matter, water, and air. Important though these discoveries were, they rinsed soil of its beautiful complexity, which has only just begun to be appreciated. In recent decades, experts have discovered that soil plays host to a living, breathing ecosystem known as the soil food web (see pp.44–45), which feeds and nurtures the plants that grow in it.

SOIL COMPONENTS

Crumbled specks of rock (called "mineral" or "inorganic" material) make up the largest part of soil and give it its permanent substance. A smaller fraction – but perhaps most important – is formed from decomposing and decayed plants and animals. This is known as "organic matter", because it is derived from organisms that were once living, and is what gives soil its chocolate brown colour. Countless soil-dwelling creatures work to gradually decompose dead leaves, roots, and other soil organisms. When they feed, digest, and poop out these remains the soil is enriched with dark, carbon- and nutrient-rich organic matter, which is what makes good garden soil feel like soft, crumbly cake.

COMPLEX STRUCTURE

Together, these mineral particles and organic matter form small, stable clumps called "peds" or "aggregates" that give soil its structure. Between these aggregates are tiny gaps ("pores"), into which air and water freely flow, creating a vast subterranean network of microscopic tunnels that are inhabited by soil-dwelling organisms and allow roots to grow unimpeded. Soil with a good structure allows water to percolate downwards through larger pores, but also retains water in smaller pores, where it can be absorbed by plant roots. These pores also mean that healthy soil is actually 25 per cent air, which is essential for roots and soil organisms to breathe. Soil is much more dynamic than it appears and the air in the top 20cm (8in) of well-drained soil is renewed hourly.

WHAT KIND OF SOIL DO I HAVE?

All soils are a combination of sand, silt, and clay. Working with your soil and watching how water drains after heavy rain is often enough to give an indication of soil type, but you can also apply the simple ball test, by taking a handful of soil and trying to roll it into a ball: sandy soil will not form a smooth ball and feels gritty when rubbed between your fingers; silty soil comes together in a ball with a silky surface that is not sticky and is difficult to roll into a long sausage shape; clay soil forms a smooth, sticky ball that rolls into a sausage and bends into a "U" shape without breaking.

SANDY SOILS

The largest mineral particles are termed sand. Soil where sand predominates is described as "free draining" or "light", because water moves downwards quickly through the relatively cavernous air spaces or pores between large sand particles. This means that nutrients can easily be washed away, but allows drier sandy soils to warm up fastest in spring. After rain, the walls of each pore are left coated with a thin sheen of water, and it is this moisture that plant roots access between waterings. Being large, sand particles have a smaller surface area for water to cling to than smaller particles which means sandy soils retain less moisture.

SILTY SOILS

Silt particles are smaller than sand, like unimaginably tiny pebbles. Silty soils are something of a halfway house between sand and clay. They hold water and nutrients better than sandy soils, while being less prone to waterlogging than clay-rich soils. Silty soil warms up faster than denser clay in spring, but slower than light sandy soil. When dry, soil that is predominantly silt becomes dusty and is easily blown away.

CLAY SOILS

The tiniest, microscopic flakes of mineral are called clay. Soils with more than 30 per cent clay tend to be be claggy when wet, yet hard when dry. This makes them difficult to dig, which is why gardeners call them "heavy", and means they warm up slowly in spring, delaying sowing and planting. Clay particles clog pores between sand and silt, impeding drainage and causing waterlogging. But they readily form larger peds with organic matter that help improve soil structure. Clay soil can retain 10–100 times more nutrients than sand and silt, giving it the potential to be fabulously fertile and allow plants to thrive.

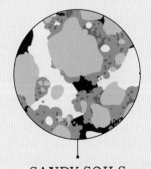
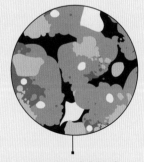
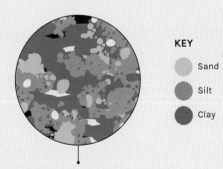

KEY
Sand
Silt
Clay

SANDY SOILS
Particles 0.2mm–2mm wide give sandy soils a coarse, gritty texture. Such large particles allow water to drain freely, but also leaves sandy soils prone to leach nutrients.

SILTY SOILS
Particles 0.002–0.02mm give silt a smooth, silky texture when wet. Small particles allow it to hold onto water and nutrients well, but leave it prone to waterlogging.

CLAY SOILS
Less than 0.002mm in size, clay flakes fill in between other particles and bind with organic matter, which can enhance nutrient retention, but risks waterlogging.

WHAT IS SOIL PH AND HOW DOES IT AFFECT WHAT I CAN GROW?

Of all the qualities in soil, its pH probably holds the greatest sway over what grows well. A measure of acidity of the water that surrounds soil particles, pH is called the "master variable" by soil scientists, because it influences all life within soil.

Soil pH varies from very acidic (4) to very alkaline (10) and, worldwide, is generally skewed to the acidic (less than 7) side of the scale. A healthy range for most plants is pH5.5–8.5, although the sweet spot for growth is pH6.2–6.8. Few sources explain that soil's acidity level is dictated by the rock that it comes from (see p.39) – usually the bedrock deep beneath topsoil – through which soil water has seeped. Soil from granite or shale is usually acidic, whereas limestone (including chalk) based soils will be alkaline (the opposite of acidic). Clay soils drain slowly and tend to retain alkaline substances, giving them a higher pH; free-draining sandy soils are generally more acidic. Long term fertilizer use and pollution also tend to acidify soil.

NUTRIENT UPTAKE

The pH of soil is important because it affects the nutrients that roots are able to absorb, regardless of the amounts that are in the soil. A high soil pH tends to make phosphorus and many trace elements (see pp.122–123) difficult for plants to take in and can cause "lime-induced chlorosis", where leaves yellow between veins because a plant cannot absorb enough iron or manganese.

In acidic soil with a low pH, calcium, phosphorus, and magnesium can be locked up, while aluminium and manganese become more available, potentially to toxic levels. All members of the soil food web (see pp.44–45) are also affected by the pH of the water in soil. For example, earthworms are most abundant where pH is close to neutral, while soil bacteria slow down as pH falls, preventing them from decomposing dead animal and plant material.

WORK WITH WHAT YOU HAVE

Each plant species evolved to cope with the pH of its natural habitat, so knowing your soil pH allows you to choose what will grow best in your patch. Lots of pH testing kits are available, but the only way to establish your soil pH with certainty is with a lab test, because a tiny inaccuracy in the result is actually a big mistake – pH is measured on a "logarithmic" scale, meaning a one-point difference equates to a ten-fold change: pH5 is ten times more acidic than pH6.

Advice on how to "correct" soil's pH abounds. Want to grow acid-loving "ericaceous" plants? Add sulphur-based fertilizer. Want healthy vegetables? "Dig in" lime. It is difficult to know exactly how much of these amendments to apply, however, and their use is ultimately fruitless, because soil water soon reverts to the tendencies of its "parent" rock. Dabbling with pH also puts the health of the soil food web at risk. The best strategy for all soil types is to regularly add a layer of compost to the soil surface ("mulching"). This tends to bring the pH towards neutral (7) while buffering against further pH shifts. Better still, simply delight in what thrives naturally in your garden. If you're really keen to grow plants that won't flourish in your soil, create ideal conditions in a container or raised bed.

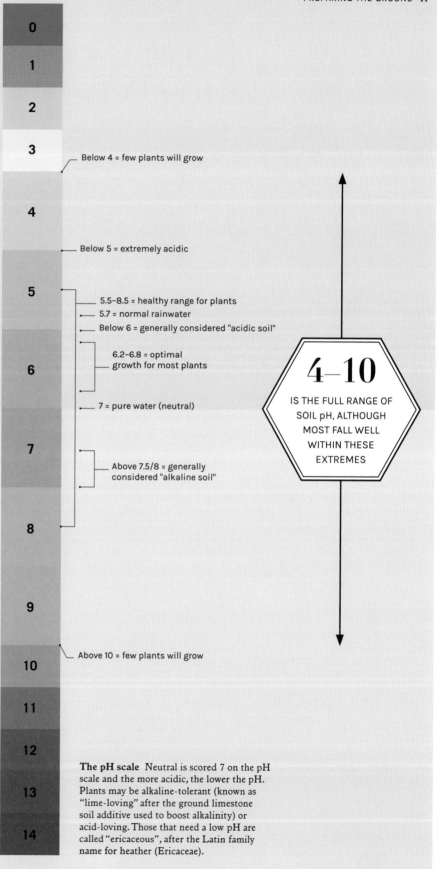

CAMELLIA
JAPANESE MAPLE (*Acer*)
BLUEBERRY (*Vaccinium*)
HEATHER (*Calluna, Erica*)
RHODODENDRON
MAGNOLIA

ACID-LOVING PLANTS
(pH<7)

ALKALINE-(LIME-) LOVING PLANTS
(pH>7)

LILAC (*Syringa*)
MARJORAM (*Origanum*)
LAVENDER (*Lavandula*)
LILY-OF-THE-VALLEY (*Convallaria*)
PHACELIA

0
1
2
3 — Below 4 = few plants will grow
4
— Below 5 = extremely acidic
5
— 5.5–8.5 = healthy range for plants
— 5.7 = normal rainwater
— Below 6 = generally considered "acidic soil"
6
— 6.2–6.8 = optimal growth for most plants
— 7 = pure water (neutral)
7
— Above 7.5/8 = generally considered "alkaline soil"
8
9
10 — Above 10 = few plants will grow
11
12
13
14

4–10
IS THE FULL RANGE OF SOIL pH, ALTHOUGH MOST FALL WELL WITHIN THESE EXTREMES

The pH scale Neutral is scored 7 on the pH scale and the more acidic, the lower the pH. Plants may be alkaline-tolerant (known as "lime-loving" after the ground limestone soil additive used to boost alkalinity) or acid-loving. Those that need a low pH are called "ericaceous", after the Latin family name for heather (Ericaceae).

HOW CAN I BEST IMPROVE MY SOIL?

Science has uprooted age-old advice that digging and adding "amendments", such as grit or lime, are the route to healthy soil. Instead it shows that soil is best improved by avoiding the spade and regularly mulching with organic matter.

Soil is the most precious thing in your garden, yet for years it has been mistreated. Digging, we are told, will remove weeds, reduce "compaction" by adding air or improving drainage, and boost soil fertility. Research now shows that the reverse is actually true.

DON'T DIG

In reality, more weeds grow after digging because buried seeds are unearthed and awoken from their dormancy by bright light and fresh air. Every stroke of a spade also destroys the structure of soil in much the same way that an earthquake will devastate the architecture of a building. Soil ends up dense, compressed, and airless ("compacted").

Healthy soil has a spongey structure that crumbles and can be pressed back together easily. Up close, organic matter and mineral material (see pp.38–39) cluster together to form "aggregates" or "peds" that give soil its structure. Glues from soil bacteria and fungi (see pp.44–45), and a sticky substance called mucilage from roots, keep these peds stable, so that tiny gaps (called "pores") form between them. Air and water freely flow through this vast network of microscopic tunnels, but this construction is left in ruins after work with a spade.

Digging creates larger pores, but only temporarily: with time and rainfall, the house crumbles down, and soil particles settle closer together, resulting in denser and more compacted soil. Worse still, this disturbance harms the delicate soil food web. Each slice with a spade severs countless fungal threads, through which plants receive nutrients and water (see pp.44–45), collapses the many tunnels forged by earthworms, and unearths sleeping plant-digesting microbes, stimulating them to feed and then release greenhouse gases into the air.

MULCH WITH ORGANIC MATTER

Protecting soil's structure and feeding its precious food web calls for mulching. As well as improving structure in all soil types, regular mulching also stops weeds growing (see pp.46–47), and helps soil retain moisture and heat, reducing the need for watering in summer, and lessening risk of frost damage in winter. Away from manicured parklands and gardens, you won't see soil left exposed in nature. That is because bare soil is prone to damage and is soon washed or blown away ("eroded") by the weather. Covering soil with woodchip, compost, straw, or rotted manure – all of which are termed "mulches" – in late autumn, protects it from pummelling winter rainfall (each bullet-like drop travels up to 30kph/20mph).

At the boundary between mulch and soil, insects, tiny bugs, earthworms, and microscopic organisms work to digest this organic material (the term for once-living matter), which is integrated into soil. Compost is made up of decomposed organic matter that is teeming with microorganisms (see pp.188–189), which rapidly feeds and adds new life to the soil food web. Another option is to keep soil covered with plants, which can even be achieved between annual plantings by sowing a "cover crop", such as clover or fenugreek. When left in place to die and decompose, these plants become "green manure", returning nutrients to the soil as they decompose.

Soil as a living system

The earth in your garden is not lifeless dust, but a dynamic ecosystem made up of diverse animal life, fungi, and bacteria, which flourish when fed and left undisturbed. Learn how to nurture it and, for your soil's sake, keep disturbace to a minimum, reserving the spade for digging planting holes.

ORGANIC MULCH
A layer of homemade or purchased compost feeds soil life, supplies plant nutrients, and protects and benefits soil structure.

WEEDS
Reduce weed growth by applying a light-excluding mulch and avoiding digging, which reveals buried seeds.

PLANT ROOTS
Healthy soil is filled with pores that provide spaces for roots to grow, and hold vital air and water.

SOIL-DWELLING CREATURES
An underground universe of life enriches soil and improves its structure, while feeding on organic matter and interacting with plants.

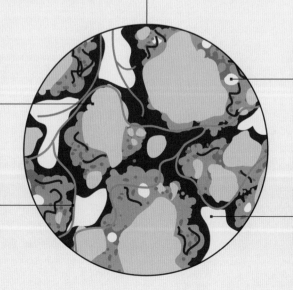

PEDS (AGGREGATES OF SOIL PARTICLES)
Minerals and organic matter are combined with sticky glues from microorganisms into larger peds that give soil its structure.

MYCORRHIZAL FUNGI
Plants form a mutually beneficial partnership with some fungi, which deliver water and nutrients to roots in exchange for energy-rich plant sugars.

SMALL PORES
These "micropores" occur in or between peds and hold onto water so that it can be accessed by roots.

LARGE PORES
Known as "macropores", these are the tunnels that allow air and water to move through soil, giving it good aeration and drainage.

WHAT IS THE SOIL FOOD WEB? (AND WHY IS IT IMPORTANT?)

Most gardeners know little about their soil, so it might come as a surprise that one teaspoonful can contain billions of living organisms. This vast community is called the "soil food web" and is vital for the health of soil, plants, and even the planet.

The 200g (7oz) of soil that you can easily hold in your hands contains 100 billion bacteria, 5,000 insects, arachnids, worms, molluscs, and minute fungal filaments that would stretch 100km (62 miles) if laid end-to-end. And that's just for starters: there are also algae, thousands of tiny worms (called nematodes), and a myriad of other microscopic creatures. All of these soil-dwelling organisms rely on one another for survival, creating an interconnecting web of life.

Single-celled organisms are eaten by larger microscopic creatures, which are eaten by worms and creepy crawlies, that are in turn food for even larger animals, such as birds. Snails, slugs, insects, and worms are just some of the invertebrates that chew, shred, and digest plant material into ever smaller pieces. All of this eating leads to poop that is food for a mass of microbes, which ultimately help turn what would otherwise be a lifeless dust into the dark, crumbly soil that plants thrive in (see pp.42–43). This mind-blowingly complex interconnected system of life is what makes soil healthy and will nourish and protect any plant that sets down roots within it.

PLANTS AND FUNGI WORK TOGETHER

Fungi play a key role in the soil food web. They are specialists at digesting what others can't, including bone, wood, tough insect shells, even rock, and are also crucial for plant survival. We may think of fungi as mushrooms, but these aboveground "fruiting bodies" are just the tip of a fungal iceberg, beneath which spread millions of tiny, hair-like "hyphae".

Mycorrhizal ("fungus-root") fungi form lasting symbiotic partnerships with about 90 per cent of plants, including almost all trees. In return for sugary fluids pumped out by roots, mycorrhizal fungi send their hyphae far further than roots can reach – sometimes for miles – in the hunt for

Hyphae inside cell

Hyphae

Fungal spore

Root cell

WITHIN CELLS

FUNGAL SHEATH

Mycorrhizal fungi can colonize roots to form a cooperative (symbiotic) relationship with plants. In exchange for plant sugars, fungal threads (hyphae) pass nutrients into roots by either tunnelling inside, or by coating them with a 'sheath' of their hyphae.

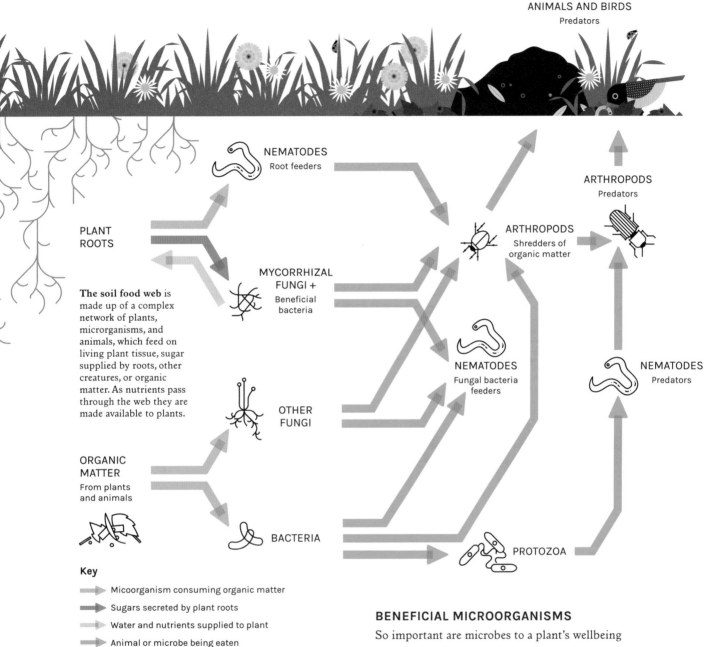

ANIMALS AND BIRDS
Predators

NEMATODES
Root feeders

ARTHROPODS
Predators

PLANT
ROOTS

MYCORRHIZAL
FUNGI +
Beneficial
bacteria

ARTHROPODS
Shredders of
organic matter

The soil food web is
made up of a complex
network of plants,
microrganisms, and
animals, which feed on
living plant tissue, sugar
supplied by roots, other
creatures, or organic
matter. As nutrients pass
through the web they are
made available to plants.

NEMATODES
Fungal bacteria
feeders

NEMATODES
Predators

OTHER
FUNGI

ORGANIC
MATTER
From plants
and animals

BACTERIA

PROTOZOA

Key

Micoorganism consuming organic matter

Sugars secreted by plant roots

Water and nutrients supplied to plant

Animal or microbe being eaten

valuable water and nutrients. These hyphae actually
penetrate the cells in plant roots (see left) and
supply a bounty of nutritious goodies, such as
nitrogen, phosphate, and zinc. Incredibly, fungal
threads also form a web to connect plants, serving
as conduits for sharing nutrients, sending chemical
alerts of disease, and even deploying toxins to
poison an enemy.

BENEFICIAL MICROORGANISMS

So important are microbes to a plant's wellbeing
that plants will use up to 40 per cent of the
energy they harvest from the sun to produce sticky
secretions in the soil to feed them. Microscopic life
swarms around plant roots, eating the sugars they
exude and offering them food and protection in
return. Some bacteria can even provide plants
with their most important nutrient – nitrogen (see
pp.122–123) – by extracting it straight from the air,
something humans can only accomplish in industrial
vats under high pressure at over 400°C (752°F).

SHOULD I GET RID OF WEEDS?

A weed can be any plant growing in the wrong place, but beauty is in the eye of the beholder. Many plants traditionally regarded as weeds are attractive and valuable to wildlife, but there are others that you're definitely best without.

———————

Weeds are great survivors that win out in the brutal battle for soil and light. Their presence may trouble you, but these much maligned plants often benefit soil and wildlife. They are fast-growing, quickly covering bare soil to shield it from erosion, and their flowers are often loved by insects. Research shows that pollinators visit some native weeds four times more often than plants in wildflower seed mixes.

There are various ways to control weeds, most of which impact on your free time rather than the environment, but although effective, synthetic weedkillers are best avoided, as research shows that they may harm the microbes and creatures within your soil. Given their benefits, it is probably time that we learned more about our weeds and found space for some of them alongside cultivated plants.

ANNUALS AND PERENNIALS

Plants labelled as weeds grow rapidly at the expense of their neighbours, hogging sunlight or even

Dandelion These perennials have vibrant flowerheads, which are an important source of spring nectar for insects, so there is good reason to leave a few in a quiet corner.

NECTAR-RICH

EACH DANDELION FLOWERHEAD IS MADE UP OF HUNDREDS OF TINY POLLEN- AND NECTAR-FILLED FLOWERLETS

- Fused petals
- Stigma and style (female flower parts)
- Anthers (male flower parts – for pollen)
- Nectary
- Pappus (seed parachute)
- Ovary

strangling them with wiry stems. Weeds are either perennials that grow back year after year from long taproots or spreading underground stems (called rhizomes or stolons), or they are fast-growing annuals which grow, flower, and shed thousands of tiny seeds in a single growing season (see below). Seeds of both annuals and perennials can blow in from neighbouring gardens, be picked up on the bottom of a shoe, or simply lie patiently in soil until they are brought to the surface to germinate (see pp.74–75).

"Invasive" plants are often imports (deliberate or accidental) from overseas, which multiply and spread rapidly, often harming local wildlife and native plants (see pp.62–63). They are always best removed as soon as they are seen.

EFFECTIVE WEED CONTROL

Pulling annual weeds up before they produce seeds will thwart the production of the next generation, but be careful to minimize soil disturbance, because there may be a store of seeds just waiting for a chance to germinate when brought to the surface. Running a hoe (a long-handled tool with a sharp blade) through the soil surface in dry weather will cut and kill weed seedlings. Perennials are harder to remove because many can regenerate from tiny pieces of root (see pp.136–137) and thorough weeding is needed to remove every scrap from soil.

A less labour-intensive way to stop weeds in their tracks is to rob them of light to prevent photosynthesis (see pp.70–71) and so deny plants the energy that is essential for their growth. Do this by completely covering bare soil (termed "mulching") with cardboard, black plastic, or any material that blocks out the sun. Even perennials growing from roots rich in stored energy will eventually die (although it may take a year or more). Better yet, an organic mulch that will rot down, such as compost or bark chippings, will have the same effect if laid at least 5cm (2in) thick, and will also nourish your soil (see pp.42–43).

A NUTRIENT SOURCE?

These greedy growers are stuffed with nitrogen and other precious nutrients (see pp.122–123), which can be returned to soil by allowing them to dry out before adding to the compost heap (see p.192) or steeping plants in water for 2–4 weeks to make "tea" to use as a liquid plant feed. Many gardening sources will tell you that deep-rooted comfrey is rich in potassium, while nettles contain magnesium, sulphur, and iron. In reality, however, the science shows us that comfrey's taproot is used for energy storage rather than to take up a specific nutrient, and weeds, just like all plants, contain varying quantities of different nutrients.

INVASIVE PLANTS

RESEARCH INDICATES THAT IT TAKES

10
YEARS

TO ERADICATE

GIANT HOGWEED
(*Heracleum mantegazzianum*)

3–4
YEARS

TO ERADICATE

JAPANESE KNOTWEED
(*Reynoutria japonica*)

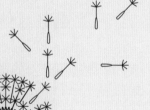

40,000
SHEPHERD'S PURSE
(*Capsella bursa-pastoris*)

25,000
CHICKWEED
(*Stellaria media*)

GROUNDSEL
(*Senecio vulgaris*)

12,000
DANDELION
(*Taraxacum officinale*)

Prolific seeders
These common garden weeds produce thousands of seeds in a matter of weeks and quickly cause problems for gardeners.

IS IT EASIER TO GROW IN CONTAINERS?

Growing plants in containers can offer advantages to gardeners and be as successful as tending a patch of land. But confining roots to a pot means plants have to rely on you for water and nutrients.

———————

Plants can thrive in any container with drainage holes in its base, from a small pot to a large raised bed. You have complete control over the potting compost used (see right), which means you can create the perfect home for plants that wouldn't thrive in your soil, for example, growing blueberries in pots of ericaceous (acidic) compost if your soil pH is too high (see pp.40–41).

SIMPLER PROTECTION

Weeds are less of an issue in sterilized compost that's free of weed seeds, and slugs and snails wreak less slimy havoc when plants are raised above the ground. Pots can also be moved under cover when cold weather threatens, meaning tender plants can be protected from frost more easily.

MORE MAINTENANCE

All this can be a mixed blessing, however. Raised above ground and with their limited volume of compost, containers are more vulnerable to freezing in cold weather. They also dry out faster than open soil and need regular watering, especially as they receive little rainfall where the compost is covered by foliage.

Confined roots potentially limit the size of plants, as well as access to nutrients, meaning that regular feeding with a suitable fertilizer (see pp.124–125) and potting on (see pp.84–85) are needed to provide the space and resources for healthy growth. Plants are also denied the company of microbes and fungi that give them nourishment and protection in soil (see pp.44–45).

Access to resources for growth

With no access to the water reserves and nutrients available in open soil, pot plants ultimately rely on you.

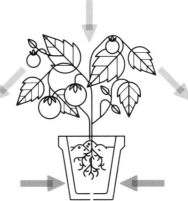

WATER
Rain falls on soil and is absorbed, making it available to roots.

ROOTS
In soil, roots seek water and nutrients, helped by microbes and fungi.

IN THE GROUND

WATER
Raindrops run off leaves and compost receives little water.

ROOTS
Inside a pot, roots can't spread in search of water and nutrients.

IN A CONTAINER

WHAT TYPE OF
COMPOST SHOULD I BUY?

A befuddling array of composts is available, each formulated for different purposes and stages of plant development. Choose those that suit your plants' needs and consider the environmental impact of their ingredients.

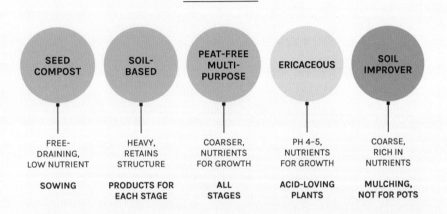

SEED COMPOST	SOIL-BASED	PEAT-FREE MULTI-PURPOSE	ERICACEOUS	SOIL IMPROVER
FREE-DRAINING, LOW NUTRIENT	HEAVY, RETAINS STRUCTURE	COARSER, NUTRIENTS FOR GROWTH	PH 4–5, NUTRIENTS FOR GROWTH	COARSE, RICH IN NUTRIENTS
SOWING	PRODUCTS FOR EACH STAGE	ALL STAGES	ACID-LOVING PLANTS	MULCHING, NOT FOR POTS

Composts used in containers are also called "potting mixes" to avoid confusion with garden compost. Choose products carefully, since the texture, drainage, nutrients, and pH of each need to match their intended use.

FIT FOR PURPOSE

Seed composts have good drainage and low nutrient levels to suit germinating seeds, but tests have shown that quality brands of multipurpose compost perform equally well for sowing. Given that multipurpose compost is also ideal for growing on seedlings and short-term plantings, it may be worth trying your own seed-growing experiments. Select specialist mixes for plants with exacting needs, such as ericaceous compost for acid-loving plants.

The ingredients in most potting mixes come from organic matter (see p.207), but some contain topsoil (termed "loam") that's heat-sterilized to kill weeds and microbes. These composts are heavier and retain a good structure for roots over several years, making them suitable for planting long-lived shrubs or trees.

GOING PEAT-FREE

Environmental concerns over the use of peat (see pp.50–51) have prompted the development of many peat-free products. Peat substitutes, such as wood fibre or coir, hold water and give bulk to compost, but are not as consistent as peat. Watch for poor quality batches that give disappointing results.

Coir

THIS BY-PRODUCT OF THE COCONUT INDUSTRY IS BILLED AS AN ENVIRONMENTALLY FRIENDLY ALTERNATIVE TO PEAT.

But it takes six months physical and chemical processing, and copious water, to turn coconut husks into coir blocks, which are then shipped long distances from India, Sri Lanka, and southeast Asia.

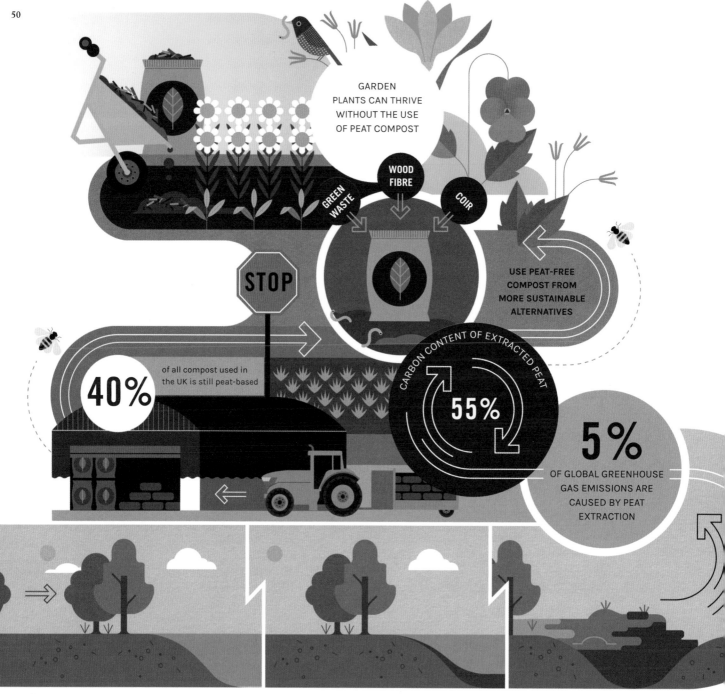

GARDEN PLANTS CAN THRIVE WITHOUT THE USE OF PEAT COMPOST

GREEN WASTE

WOOD FIBRE

COIR

USE PEAT-FREE COMPOST FROM MORE SUSTAINABLE ALTERNATIVES

STOP

40% of all compost used in the UK is still peat-based

CARBON CONTENT OF EXTRACTED PEAT

55%

5% OF GLOBAL GREENHOUSE GAS EMISSIONS ARE CAUSED BY PEAT EXTRACTION

PEAT IS THE PRODUCT OF PLANT MATERIAL DECOMPOSING IN A WETLAND ENVIRONMENT

ACIDIC CONDITIONS AND A LACK OF AIR IMPEDE DECOMPOSITION

THIS RESULTS IN THE FORMATION OF PEAT BOGS OVER MILLENNIA

From bog to garden

The story of peat is thousands of years in the making. By using it in our gardens, we contribute to the destruction of ancient habitats and the release of climate change gases into the atmosphere.

44% OF THE TOTAL GLOBAL SOIL CARBON IS LOCKED INSIDE PEAT

3% OF GLOBAL LAND MASS IS COMPRISED OF PEATLAND

WHAT'S SO BAD ABOUT USING PEAT?

Peat can seem like brown gold to a gardener, with properties that can make even the fussiest of plants bloom with delight. This wonder substance is still used in most potting mixes, but should really be consigned to history: it is like burning coal to keep your petunias pretty.

THE BENEFITS OF PEAT

Peat holds on to water better than most sponges, gives structure and body to soil better than any mineral or man-made substance, it doesn't decay or wash away quickly, and gives growing roots a firm foothold. Each slice of the dark spongy earth is as identical as the last, making it far more consistent and reliable than any compost or topsoil. Under a microscope, peat looks like honeycomb, with each cavity of soft carbon offering a home to plant-friendly bacteria and a perfect place for holding soil nutrients.

RAZING PREHISTORIC FORESTS

A block of peat contains thousands of years' worth of compressed woodland and vegetation, all partly decomposed. The act of breaking it up and crumbling it into potting mixes and composts exposes it to the air, awakening plant-munching microbes that had lain dormant in the dark for centuries. Now released from their acidic paralysis, they belch out invisible clouds of locked-away carbon dioxide and methane. It is as if you are burning prehistoric forests in your back garden.

Peat is so rich in compressed carbon that for centuries we have burnt dried bricks of peat to heat our homes and provide energy, and still do in some countries today. Peat is a profoundly dirty fuel, sometimes dubbed the "forgotten fossil fuel" and remarkably, peatlands contain more carbon than all the world's forests combined.

PEATLANDS LAID WASTE

Gardeners who use peat are also contributing to the destruction of an irreplaceable habitat. Unlike regular soils, the bogs and marshes that make up peatlands are so utterly waterlogged that the fungi, bacteria, and critters that normally break down leaves, twigs, and other fallen vegetation are suffocated. The earth becomes an acidic, half-rotten swamp. It should be a wasteland, but nature finds a way and peatlands are carpeted with unusual flora and mosses (particularly sphagnum moss) and play host to many rare, threatened, and declining plants and animals. It is a gut-wrenching irony that nature lovers would carve up such precious areas of biodiversity for mere horticultural convenience.

A "sustainable" option?

SOME PEAT PRODUCTS ARE MARKETED AS SUSTAINABLE, BUT SUCH CLAIMS HAVE BEEN LARGELY DISCREDITED BY SCIENTISTS.

Peatlands are replaced at a rate of 1mm each year as surface moss grows and dies, continually adding its remains to the wet earth. Some North American peat producers sell so-called sustainable peat by claiming that they only strip off the dry layer underneath the surface moss and allow it to fully regrow before gathering more, or that the amount they take is a tiny proportion of the whole. Conservationists say no peat product is sustainable, and for every handful of any kind of peat harvesting, plumes of carbon dioxide are released into the air.

DO I NEED TO
USE PESTICIDES?

Sprays offer a quick fix for weeds or aphids, but their ingredients can be harmful to humans, the environment, and may even worsen problems they claim to solve. Reducing pesticide use has many benefits and better alternatives are available.

Pesticide is a catch-all term for any substance used to kill living things that are deemed "pests". It includes chemicals to kill weeds (herbicides), insects (insecticides), fungi (fungicides), rats and mice (rodenticides), as well as slugs and snails (molluscicides). By their very nature, all pesticides are poisonous to life, but some have the potential to be more harmful than others.

Organic products are based on naturally occurring substances which, although they are still hazardous, are biodegradable and break down within hours or days. Synthetic pesticides contain man-made active ingredients that can linger on plants or in soil for weeks, months, or even years (termed "persistent pesticides"), giving them much greater potential to

harm visiting insects or creatures living in soil (see pp.44–45). Pesticides can also seep away through soil, ending up in streams and waterways, where they can poison aquatic life. Today's pesticides are highly regulated and go through extensive testing, but nevertheless can have unintended consequences: highly effective neonicotinoid insecticides poison the brains of bees; while widely used glyphosate herbicides affect the health of microbes and mammals alike, and have also been linked to cancer.

A LIFE WITHOUT PESTICIDES?

In the US, homeowners use up to ten times more pesticides per acre on their lawns than farmers do on their crops. In the light of evidence implicating

Integrated pest management

Also known as IPM, this approach offers gardeners a range of alternatives to try. It simply involves finding out which pests might plague each plant and then employing a variety of targeted strategies in a stepwise fashion to prevent or resolve any problem you identify.

 RESEARCH

Find out about potential pests and when they might strike:

● Consult gardening books and websites to identify your plants and their associated pests.

● Use scent and sticky traps to detect the presence of pests.

 CULTURAL

Use gardening techniques to prevent problems:

● Provide good growing conditions by siting the right plant in the right place (see pp.58–59).

● Add organic matter to soil for healthy plants with strong natural defences.

● Rotate crops and tidy up diseased plant material to prevent problems occurring, persisting, and spreading.

pesticides in plummeting insect populations, as well as concerns over them harming our health, many gardeners are reconsidering whether they are really worth it to prevent a few weeds, nibbled leaves, or imperfect fruit.

If given the chance, natural predators, such as birds and ladybirds, will come to the rescue and feast on insect pests like aphids. Dousing plants with insecticide at the first sign of trouble not only wipes out this food source for wildlife, but may kill the insect predators too, meaning they won't be there to control aphid numbers when they inevitably return. Without the interference of pesticides, populations of pests and predators can reach a natural balance, where some pests are always present, but at tolerable levels that remain controlled – gardening without pesticides means learning to accept their presence. Interspersing prized plants with a variety of others that attract beneficial creatures (see pp.140–141) can help to achieve this balance.

TREATMENT TERMS
UNDERSTANDING THE WORDS USED ON PESTICIDE LABELS HELPS YOU CHOOSE A SUITABLE PRODUCT.

CONTACT
Acts only upon what it touches: killing leaves not roots, or only insects that are hit with a spray.

SYSTEMIC
Absorbed by plants and distributed through all tissue to kill the whole plant or pests that feed on it.

SELECTIVE
Targets only a particular type of plant or animal and leaves others unharmed, for example, lawn weedkiller.

NON-SELECTIVE
Kills both pests and beneficial bugs, or all weeds and garden plants indiscriminately.

PHYSICAL

Employ simple "physical" methods of protection:

- Keep pests away from plants using crop covers (fleece or insect mesh).
- Pick pests off plants by hand or spray off with water.

PREDATORS

Encourage wildlife that preys on plant pests:

- Grow plants that attract beneficial insects, like aphid-eating hoverflies.
- Provide trees and shrubs for birds, or even a pond for frogs and toads.
- Release natural predators to control specific pests.

PLANT INVIGORATORS

Try "natural" pesticides if damage can't be tolerated:

- Use products containing naturally-derived surfactants or oils to kill insect pests on contact.
- Diatomaceous earth is a sharp, powdered rock that deters pests.

CHEMICALS

Chemical pesticides can be used as a last resort.

- You can choose not to use pesticides if you prefer.
- Find a product appropriate for the plants to be treated and that is less likely to cause damage to wildlife.
- Apply and store according to label instructions.

IS IT HARDER TO GARDEN ORGANICALLY?

Organic gardening is more than just avoiding the use of manufactured chemicals. It is a shift in mindset that allows the plants, soil, and creatures in your plot to work together. It sounds a big ask, but in the longer term often makes gardening easier.

———————

Born in the post-war period as a backlash to intensive farming practices, organic growing is about trying to work with nature rather than conquer it. Achieving this and reducing your use of pesticides and fertilizers is simpler and less stressful when you are not striving for the "perfection" of a pristine lawn or immaculate rose beds. Look around and you'll see that nature is healthiest in all its variety, while a "monocrop" of one plant type is a honeypot for pests and diseases.

Whether cultivating vegetables or flowers, organic gardeners seek a more natural balance by growing a range of plants to provide food and shelter for the beneficial insects that pollinate plants and prey on pests. This variety also confuses pests, and if a problem does arise, it's unlikely to affect every plant. Taking advantage of the ways that plant species physically and chemically interact with one another is called "companion planting" and although this is a subject littered with folklore, science has found plant combinations that are beneficial (see p.94).

ALTERNATIVES TO PESTICIDES

Growing a combination of plants also makes the presence of a few leaf spots or aphids less obvious and easier to tolerate, which reduces the urge to reach for a spray. In fact, carpet bombing an area with a pesticide or herbicide (see pp.52–53) ultimately helps no-one, because garden plants, birds, insects, mammals, and microbes all rely on one another. An insecticide (whether synthetic or organic) intended to kill aphids can also wipe out their natural predators, such as hoverflies and ladybirds, so they are not there to feed on aphids and control their numbers when they inevitably return. Allowing some pests to persist on your plot means that their predators will be nearby to keep them in check naturally. Finding out when pests are troublesome allows you to time sowings and cover crops to prevent problems, or introduce natural predators as part of an "integrated pest management" approach (see pp.52–53), which offers lots of useful options before pesticides are considered.

Another cornerstone of organic gardening is to focus on steady growth and so produce sturdy plants, which are less likely to be attacked or seriously damaged by pests and disease. This begins by selecting plants that are suited to your soil, climate, and growing space (see pp.58–59), which will flourish with little attention.

The "No Dig" approach

"NO DIG" GARDENING MEANS LESS WORK AND BETTER HEALTH AND STRUCTURE TO YOUR SOIL. Crucial to "No Dig" is feeding soil with an annual mulch (covering) of compost, which is left on the surface and not dug in. This enriches with nutrients, improves structure, and sustains the soil food web (see pp.42–45) – plus it smothers weeds before they grow. Kept healthy in this way, soil provides plants with all the nutrients they need, removing the need for fertilizers. Starting a compost pile will yield a free supply of this black gold from kitchen and garden waste (see pp.190–191).

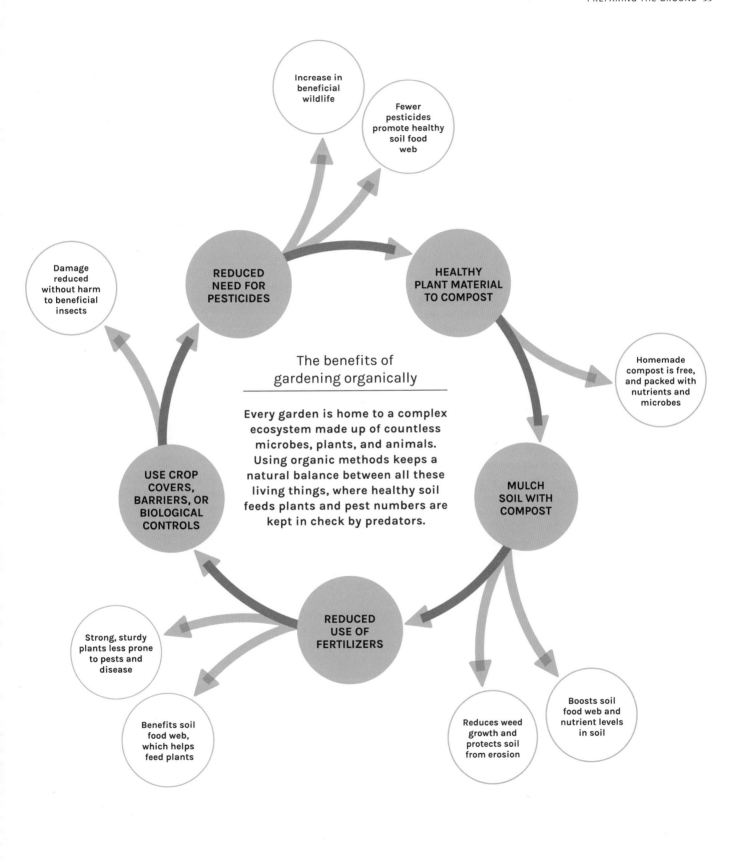

Increase in beneficial wildlife

Fewer pesticides promote healthy soil food web

Damage reduced without harm to beneficial insects

REDUCED NEED FOR PESTICIDES

HEALTHY PLANT MATERIAL TO COMPOST

Homemade compost is free, and packed with nutrients and microbes

The benefits of gardening organically

Every garden is home to a complex ecosystem made up of countless microbes, plants, and animals. Using organic methods keeps a natural balance between all these living things, where healthy soil feeds plants and pest numbers are kept in check by predators.

USE CROP COVERS, BARRIERS, OR BIOLOGICAL CONTROLS

MULCH SOIL WITH COMPOST

Strong, sturdy plants less prone to pests and disease

REDUCED USE OF FERTILIZERS

Boosts soil food web and nutrient levels in soil

Benefits soil food web, which helps feed plants

Reduces weed growth and protects soil from erosion

High versus low maintenance

Simple changes can make a huge difference to the amount of routine care your garden needs. Swap practices that generate work (left) for those that demand less effort (right).

MOWN LAWN

Lawns need regular mowing and edging, which takes time and effort.

LONGER GRASS

A long meadow-like area reduces mowing and benefits nature.

BARE SOIL

Weeds fill any free ground. Uncovered soil always needs more weeding.

GROUND COVER

Fill beds with plants so that there is no space for weeds to grow.

DIGGING

Digging in compost damages the soil structure and promotes weeds.

NO DIG

A compost mulch suppresses weeds and feeds soil, so makes less work.

WRONG PLACE

In unsuitable conditions plants struggle and will need more care.

RIGHT PLACE

Given a spot that fits their needs plants thrive with little attention.

PLANTING IN CONTAINERS

Plants in pots need frequent watering and feeding when in growth.

PLANTING IN THE SOIL

Plants are less reliant on you when roots can grow in open soil.

WHAT MAKES A GARDEN LOWER MAINTENANCE?

While no garden is maintenance-free, the upkeep of a vibrant green space need not be a chore if it is tailored to your interests and needs. Creating a space that works for you will make looking after it something you enjoy.

Many people think that having a garden means hard work. Between 2005 and 2015, the number of front gardens that were paved over in the UK increased by 200 per cent, leaving a third of homes with no planted outdoor space. But adapting what you grow, and where and how you grow it, can make gardening less time consuming and more rewarding.

LOSE SOME LAWN

Getting rid of some or all of your lawn is the fastest time saver. One square metre of lawn contains 100,000 thirsty blades of grass: few other plants need as much watering and feeding, and then there's the regular mowing. This is a lot of effort if a carpet of grass serves no practical purpose, and could be swapped for colourful planting that needs less regular care. Replacing a lawn with paving or artificial turf may be appealing, but often extinguishes plant, soil, and insect life, while amplifying flooding risks in your area (see p.23).

KEEP SOIL FULL

Avoid leaving soil bare, because it will quickly be colonized by weeds. Exposed soil will also be quickly eroded and damaged by the elements (see p.42). The "no dig" approach, where strenuous digging is replaced by mulching soil annually with compost (see p.42–43), is kinder on your back, while improving soil health and reducing the need for weeding and watering.

PLANT CHOICES

Careful plant selection will save headaches further down the line. Match plants to locations in the garden where the soil and light conditions suit their needs (see p.58–59) and they will need minimal watering and care once established. Annual bedding plants that need to be replaced once or twice a year are best avoided. Instead, opt for trees, shrubs, and perennials that grow year-after-year. Plant slow-growing hedging, such as yew (*Taxus*), that can be cut once a year and its upkeep is no more onerous than painting a fence.

Plants in containers need frequent watering (perhaps daily in summer) and regular feeding (see pp.110–111) and may be best avoided to reduce your workload in the garden.

ROOM TO GROW

CHECK HOW LARGE TREES AND SHRUBS WILL GROW; GIVING THEM SPACE WILL AVOID PRUNING LATER

WHAT DOES "RIGHT PLANT RIGHT PLACE" MEAN?

Within every garden, different growing conditions offer ideal homes for different plant species. "Right plant right place" means choosing plants to suit their surroundings, making life easier for both them and you.

It's all too easy to buy a plant on impulse, only to find that it doesn't flourish. With no idea about the growing conditions plants need, or those that exist within your garden, planting is a lottery that's likely to lead to disappointing results. Forced out of its comfort zone, a plant may struggle to get the light, water, and nutrients it needs, leaving it weak, vulnerable to pests and disease, and more reliant on you.

PERFECT PLANTS FOR YOUR PLOT

Success hinges on finding plants that come from environments similar to those in your garden. Plants have evolved to grow in almost every corner of the globe, no matter how inhospitable, and even the boggiest, shadiest corner will perfectly suit at least a few plant species. This sounds daunting, but books, websites, and garden centre staff are great sources of information and with a little research you will find that you are spoilt for choice.

Finding plants that flourish in different areas is a journey of discovery, involving some trial and error. Reduce this by finding out your garden's hardiness zone (see pp.80–81) and aspect (see pp.32–33), and matching plants to individual microclimates (see pp.36–37).

LOOK AT LIGHT AND SHADE

Light requirements are important and plants are broadly divided into "shade-loving" and "sun-loving" categories. Sun-loving plants like to be bathed in sunlight for the entire day and have evolved leaves with sun protection and water

preservation strategies built in (see p.117). Shade-loving plants have leaves capable of highly efficient photosynthesis in low light levels, but often lack biological sunscreen, so they are damaged by the full force of the sun.

Many plants fall between these extremes and are best with just four to six hours of sun each day – ideally in the morning, when it is cooler and less harmful. Sunlight filtered through leafy branches (termed "dappled") is often recommended for woodland plants, but there is no evidence that this suits them better than any partial shade.

ASSESS YOUR SOIL

Soil conditions are also crucial and may even vary within a few metres, so it's vital to check the soil type (see p.39) and note if it's predominantly dry, fast-draining sandy soil, which suits many Mediterranean plants, or contains more clay, where moisture-loving plants will flourish.

Soil pH (see pp.40–41) is also critical, affecting the ability of plants to extract some nutrients from the soil, which means it's useful to test soil pH before planting. By peering into local gardens you will glean clues to the plants that will likely thrive.

Matching plants to places

Differences in light, shade, temperature, wind, and soil conditions will all impact growth. Select plants carefully to suit each part of your garden.

WET AND FULL SUN

Soil is damp around the south-facing pond, which is sunny all day.

MOIST AND PART SHADE

A north-east facing fence only gets cool morning sun so soil retains moisture.

DRY AND SHADE

The free-draining raised bed tends to be dry and is in north-facing shade.

MOIST AND SHADE

A pergola creates shade and clay-rich soil remains consistently moist.

DRY AND FULL SUN

This south-facing wall takes the sun's full force and the soil receives little rain.

GARDEN ROOM / OFFICE

PATIO AREA

FRONT DOOR

DRY, FULL SUN	WET, FULL SUN	MOIST, PART SHADE	DRY SHADE	MOIST SHADE
Need free-draining soil and an open or south-facing position.	Require wet conditions, in bog gardens or clay soils, and a bright spot.	Thrive in soil that holds moisture and shaded from strong sun.	Cope with little water and full shade; ideal under shrubs or trees.	These lush plants need good soil and shading from harsh sunlight.
ROSEMARY (Rosmarinus)	Ligularia 'The Rocket'	HELLEBORE (Helleborus)	BARRENWORT (Epimedium)	JAPANESE MAPLE (Acer)
AFRICAN LILY (Agapanthus)	CANDELABRA PRIMROSE (Primula bulleyana)	JAPANESE ANEMONE (Anemone x hybrida)	BUGLE (Ajuga reptans)	RHODODENDRONS
CATMINT (Nepeta)	YELLOW FLAG IRIS (Iris pseudacorus)	HEUCHERAS	MALE FERN (Dryopteris filix-mas)	SNOWDROP (Galanthus)
ALLIUMS				HOSTAS

SHOULD I HAVE A LAWN?

Traditionally a green, clipped lawn has taken pride of place at the centre of a garden. Over recent years, however, our obsession with owning a square of turf has been called into question, as the environmental costs of maintaining it have become clear.

The modern lawn began in late 12th-century England, as an area of finely cut grass where the gentry could play bowls. Ever since, owning a patch of manicured grass has been a marker of prestige and, after the lawn mower hit the shops in the late 1800s, the lawn became a symbol of success for any self-respecting homeowner. But is it an essential part of a 21st-century garden?

THE TROUBLE WITH TURF

This emerald green perfection that gardeners strive for comes at a price. Lawns are a monocrop of just one plant type that is prevented from flowering or producing seeds by frequent mowing. Short turf is a wilderness for most wildlife, offering sparse shelter and no food for insect pollinators. Turf grasses are shallow rooting and can only access water close to the soil surface, so need incessant watering in hot, dry weather to keep them green – although they actually survive

dry periods well and green-up as soon as rain falls. Add to that the harmful weedkillers, fertilizers, and energy-hungry mowing required for a pristine, weed-free lawn and this "must-have" starts to look like an environmental disaster zone.

MOVING AWAY FROM A BOWLING GREEN

A garden lawn can be a joy, however, and there are ways to make it less of a burden on the planet. When sowing or laying turf, opt for a harder-wearing mix of grass species that will need less care than a fine lawn. Some seed mixes also contain tiny clovers which reduce the need for feeding, by virtue of clover's ability to harvest nitrogen from the air (see p.122). Although not as good as trees for capturing greenhouse gases, lawns where clippings are left to decompose pull CO_2 from the air and are able to hold some carbon in soil.

Allowing native plants that are often considered weeds to flourish in your lawn can be a

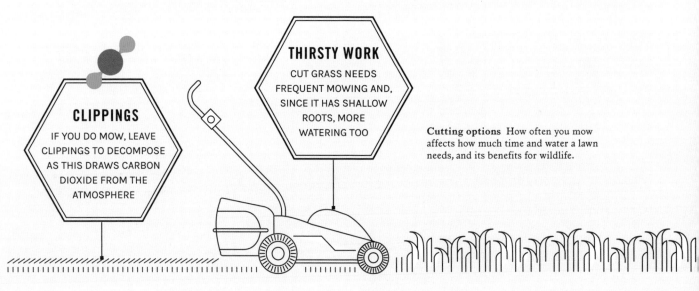

CLIPPINGS
IF YOU DO MOW, LEAVE CLIPPINGS TO DECOMPOSE AS THIS DRAWS CARBON DIOXIDE FROM THE ATMOSPHERE

THIRSTY WORK
CUT GRASS NEEDS FREQUENT MOWING AND, SINCE IT HAS SHALLOW ROOTS, MORE WATERING TOO

Cutting options How often you mow affects how much time and water a lawn needs, and its benefits for wildlife.

Lawn choices

Consider whether a mown lawn is the right option to give you an attractive garden that's easy and sustainable to maintain.

CLIMATE	PRACTICAL NEEDS	TIME TO MOW	CHEMICAL USE	WILDLIFE
Does the weather suit grass? Water is needed to keep lawns green in hot, dry summers.	**How will you use a lawn?** Do you need space to play or could it be used differently?	**Can you maintain a lawn?** Regular mowing and edging is time consuming.	**Are fertilizers and pesticides necessary?** A green lawn is possible without them.	**Can your lawn be a resource for wildlife?** Including flowers and mowing less helps.

difficult step when you've tried to eradicate them. But many, including daisies (*Bellis perennis*) and white clover (*Trifolium repens*), are capable of flowering in a mown lawn, adding attractive colour during spring and summer, and a valuable resource for insects. The decision to include them instantly removes the need to use harmful weedkillers, which can damage other plants and the creatures that make up the soil food web (see pp.44–45).

These, and the high-nitrogen fertilizers used to feed and green-up lawns, can also pollute ground water.

Avoiding a petrol mower (see pp.30–31) and being more relaxed about cutting will also reduce energy use, air pollution, and do wonders for wildlife. Leaving even part of the lawn unmown allows grasses a chance to set seed and nectar-rich plants to flower, creating a rich food source for many creatures.

LEFT LONG
LAWN LEFT UNCUT IN LATE SPRING REDUCES MOWING AND IS OFTEN FILLED WITH FLOWERS FOR INSECTS

IN THE US TODAY, **LAWN GRASS** IS THE SINGLE LARGEST IRRIGATED CROP, TAKING UP

3 TIMES

THE AREA OF THAT PLANTED WITH **MAIZE**

LAWNS FED WITH **NITROGEN FERTILIZER** EMIT

5–6

TIMES MORE CO_2 THAN THEY ABSORB DURING PHOTOSYNTHESIS

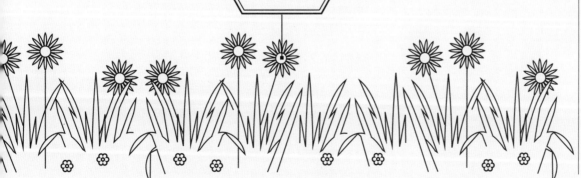

SHOULD I GROW ONLY NATIVE PLANTS?

Native plants provide food for the wildlife they have evolved alongside, but research shows that, while occasionally non-native garden plants escape to become "invasive", many are great for wildlife and should be welcome in our gardens too.

Native plants are those that have lived in a particular region for thousands of years and have evolved alongside the local wildlife. During this time, plants and wildlife often develop special relationships, meaning that the survival of one is threatened without the other. Native plants are a valuable addition to gardens, because some flower at the perfect time to deliver nectar for particular insect pollinators, and others can even be the only food source for insects at particular stages in their life cycles, especially the larval stage (caterpillars).

Traditionally, however, gardens have been curated displays of plant life, boasting specimens hand picked from around the world, grown in neat arrangements, and sometimes bred to be unnaturally beautiful. Of course attractive native plants have long been a popular part of these displays, but gardeners have always been captivated by plants that are new, bigger, and better. This desire for ever-larger and more extravagant blooms led plant breeders to cultivate varieties that are a feast for our eyes, but a barren distraction for insects. "Double" flowers (see pp.140–141), for example, are astonishingly beautiful but have had their pollen-bearing stalks (stamen) and nectar-producing glands (nectaries) swapped out for extra petals.

ALIEN INVADERS

Wherever humans tread, plants hitch a ride with them – deliberately or accidentally – to become a "non-native" or "alien" species in a new land. Such newcomers often only survive when grown in gardens, but the few that are able to thrive in the wild ("naturalize") can

Macedonian scabious (*Knautia macedonica*) is a native of southeastern Europe but, wherever it is planted, its pretty pincushion flowerheads make plentiful nectar available to insects.

NECTAR-RICH

KNAUTIA FLOWERS ARE MADE UP OF MANY FLOWERLETS EACH WITH ITS OWN NECTARY: A GREAT RESOURCE FOR INSECTS

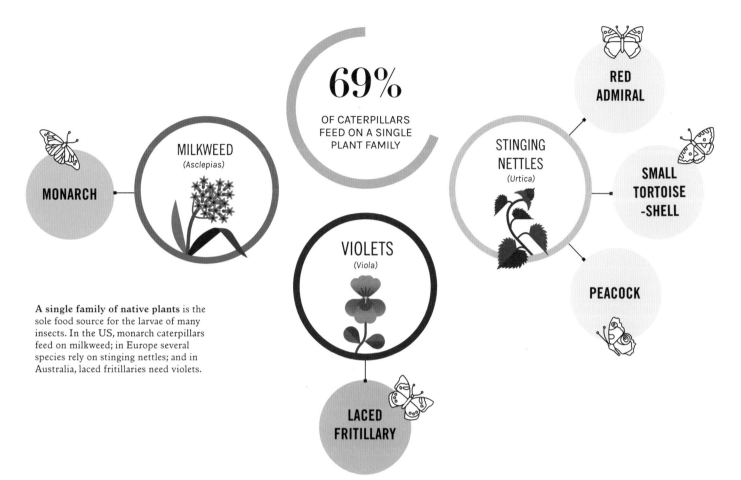

69%

OF CATERPILLARS
FEED ON A SINGLE
PLANT FAMILY

MONARCH

MILKWEED
(Asclepias)

VIOLETS
(Viola)

LACED
FRITILLARY

STINGING
NETTLES
(Urtica)

RED
ADMIRAL

SMALL
TORTOISE
-SHELL

PEACOCK

A single family of native plants is the sole food source for the larvae of many insects. In the US, monarch caterpillars feed on milkweed; in Europe several species rely on stinging nettles; and in Australia, laced fritillaries need violets.

occasionally upset the apple cart, tipping a finely tuned ecosystem into a spiral of disorder by damaging the habitat or smothering native plants. It is then termed an "invasive" species, and may offer little benefit to local insect life, while pushing out the plants they rely on. To protect native plants and wildlife, check lists of invasive plants in your country, state, or region – avoid planting listed species and remove them if they're already in your garden.

A PRAGMATIC APPROACH

"Non-native" doesn't have to mean "bad", however. Fewer than one in a thousand introduced species becomes invasive, and roughly a third of all plants in the wild are already international in origin. Many imported plants are loved by gardeners and animals in equal measure – their striking blooms often brimming with pollen and nectar. Research actually shows that gardens with a mixture of native and non-native plants are best – supporting the widest variety of insect life. The definition of native can also vary: in North America, any plants present before European settlers arrived are considered native, while elsewhere, the cut-off is the end of the last ice age. Poppies, nettles, and dandelions are ancient incomers that could once have been considered invasive, but are now part of the flora in many countries. Most alien plants find a place in the natural pecking order in the same way. Always avoid known invasive plants, but don't feel bad about enjoying exotic species alongside natives that will feed local birds and insects.

INVASIVE PLANTS

JAPANESE KNOTWEED
(Reynoutria japonica)
EUROPE AND USA

HIMALAYAN BALSAM
(Impatiens glandulifera)
EUROPE

ENGLISH IVY
(Hedera helix)
USA

SOUTH AMERICAN CARPETWEED
(Phyla nodiflora)
AUSTRALIA

WHAT'S WITH ALL THIS LATIN?

Gardeners have a reputation as a friendly bunch, happy to share their knowledge and time, as well as plants. Why then, do they insist on using complicated Latin plant names that befuddle new gardeners when simpler common names exist?

Getting a handle on Latin names may seem tricky, but they are vital to prevent gardeners talking at cross-purposes. Without them, it's impossible to be sure that two people are talking about the same plant. There are many examples where "common names" can easily become muddled, especially where different languages come into play. A lime tree in the UK is what Americans know as a linden tree, while what a Spaniard calls a lime tree is an elder tree in Ecuador (and vice-versa) – the potential for confusion is huge!

SCIENTIFIC NAMES ARE SIMPLE

Today's two-name (binomial) Latin system for naming plants was a stroke of genius by Swedish botanist Carl Linnaeus, who condensed baffling 18th-century scientific plant titles into a simple two-word species name. All plants with that unique name will look alike, be genetically similar, and can breed with one another. This first two-word part of a Latin name also gives information on how species are related to one another.

Understanding the binomial system is as easy as grasping how cars are named, using make,

model, and an optional version. A Ford Mustang GT, for example, is a car made by Ford, called a Mustang, and is a GT version. Every car with this name is built according to the same blueprints and looks similar, and if you wanted to buy one you would know what to ask for. Plant names work in the same way, just using botanical groups instead: the genus in place of a manufacturer, the species name as the model, and sometimes an additional name to describe variations within a species.

GRASPING BOTANICAL GROUPS

A genus is a group of related plants to which the species belongs. Just as all cars from a manufacturer have similarities, all plants in a genus have common features and will be genetically similar. For example, the buttercup genus, *Ranunculus*, contains hundreds of species, most of which have five-petalled flowers in shades of yellow or white. Gardeners often refer to plants only by their genus, as they all have much in common.

Each distinct species within a genus is named, just like the model of a car. The names used often say something about the appearance or preferred habitat of the plant, for example, *aquatilis* means "from water", *lutea* means "yellow-coloured".

Sometimes an extra word is used to denote a variation within the species, like a special version of a car (the GT). This could be a naturally occurring variety or subspecies, or a variety selected or produced by a plant breeder, which is termed a "cultivar" and given a name in single quotation marks, chosen to suit its colour or character.

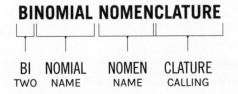

BINOMIAL NOMENCLATURE

BI	NOMIAL	NOMEN	CLATURE
TWO	NAME	NAME	CALLING

A unique label Two-word Latin names are used universally, making it possible to trade, study, and discuss plants internationally.

Naming system

Latin names always follow the same pattern, with a plant's genus and species first, sometimes followed by a further name (or names) to denote differences within a species.

SUBSPECIES (SUBSP.)
A NATURAL VARIATION FROM THE SPECIES IN ONE OR MORE WAYS

FORM (F.)
DIFFERS FROM THE SPECIES IN ONE MINOR WAY, SUCH AS FLOWER COLOUR

VARIETY (VAR.)
INDICATES A SLIGHT DIFFERENCE IN BOTANICAL STRUCTURE FROM THE SPECIES

Genus   *species*   (..............)   'Cultivar'

a group of plants with distinctive features in common that are related

Shown as
Capitaliized, in italics, may be abbreviated to first letter

plants within a genus that reproduce naturally, forming offspring similar to the parents

Shown as
Lower case and in italic

a variety bred or selected for useful or decorative garden qualities

Shown as
Capital letter, Roman type, and in single quotes marks

Geraniums (cranesbills) are a genus of plants with similar features, that would be easily confused without Latin names. They can be annual, biennial, or perennial, and vary in size and preferred growing conditions.

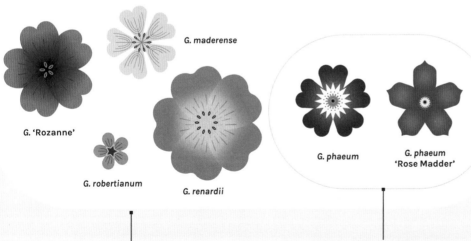

G. maderense

G. 'Rozanne'

G. robertianum

G. renardii

G. phaeum

G. phaeum 'Rose Madder'

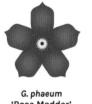

G. phaeum 'Rose Madder'

Geranium

This genus has over 400, mostly sun-loving species, with 5-petalled, white, pink, or purple flowers.

Geranium phaeum

This species is shade-loving and bears small, dark purple flowers on tall stems above its foliage.

Geranium phaeum
'Rose Madder'

This cultivar differs from the species in its smaller dusky pink blooms and pale green leaves.

ARE SOME GARDENING TOOLS BETTER THAN OTHERS?

Buying cheap tools is a false economy, since they're unlikely to stand the test of time or to be effective and comfortable to use. It is wise, therefore, to buy only essential tools and to invest in quality.

You only need a handful of tools to garden: a garden spade and fork for working soil; a hand trowel and fork for weeding and planting; secateurs for pruning; and perhaps a hoe to dislodge weeds and a rake to level soil. Before buying, handle any tool to make sure it is easy to grip and correctly proportioned for your height.

ENGINEERED TO LAST

The garden spade is used for moving soil and planting, and isn't all that different to digging sticks used millennia ago: a long sturdy shaft with a handle, for reaching and levering, and a narrow blade for cutting into soil. Forks have sharp points (tines), which can more easily pierce and loosen hard soil,

and dense garden compost. Tests show that wooden shafts perform better than heavier metal – ash wood is prized for its durability and shock resistance. A spade or fork blade will lever soil with more than double the pushing force you put on the shaft, meaning it is important that the blade is reinforced with a long collar (the metal extension of the blade that covers the shaft) and sturdy rivets.

WEED WITH EASE

Hoes have a head with a sharp blade that is angled to cut through the top layer of soil and slice through weeds. Although hand hoes are available, long-handled versions are more comfortable to use when weeding large areas. Blade design varies: draw

Choosing cutting tools

Good secateurs and loppers make clean, precise pruning cuts (see pp.170–171). They have a sharp top cutting blade and a blunt lower blade, which the stem is pushed against. "Anvil" and "bypass" designs suit different pruning tasks. Bypass blades are best for most pruning.

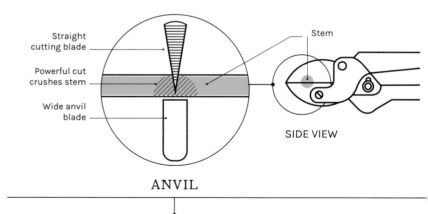

Straight cutting blade

Powerful cut crushes stem

Wide anvil blade

Stem

SIDE VIEW

ANVIL

Both blades are straight-edged. The sharpened edge of the top blade strikes a central groove in the bottom anvil blade. This powerful sandwiching action is inaccurate, and crushes the cut stem. Use to cut dead or diseased wood.

hoes have a blade angled downwards that should
be pulled towards you; Dutch hoes have a shallow
angle and are pushed away from the body; stirrup
(or loop) hoes have a pivoted blade sharpened on
both sides, which can cut efficiently on both the
push and pull stroke.

CHOICE OF METAL MATTERS

The metal used to make the blade of any tool
will affect its performance and the maintenance
it needs. Stainless steel blades don't rust, stay
shiny with little cleaning or other maintenance,
and perform best in tests of digging tools. Carbon
steel (pure steel without anti-rust additives) is
durable, but quickly rusts, so blades need to be
cleaned and oiled after use. Copper is a practical
and attractive metal for gardening tools, but it is
slightly heavier than steel and expensive. Claims
that copper tools enrich the soil and don't disrupt
the earth's magnetic fields are make-believe.

The blades of hoes and cutting tools, such
as secateurs, need to be kept sharp. Although
stainless steel doesn't rust, it's harder to sharpen
than carbon steel, but the latter needs cleaning
and oiling after every use to prevent corrosion.

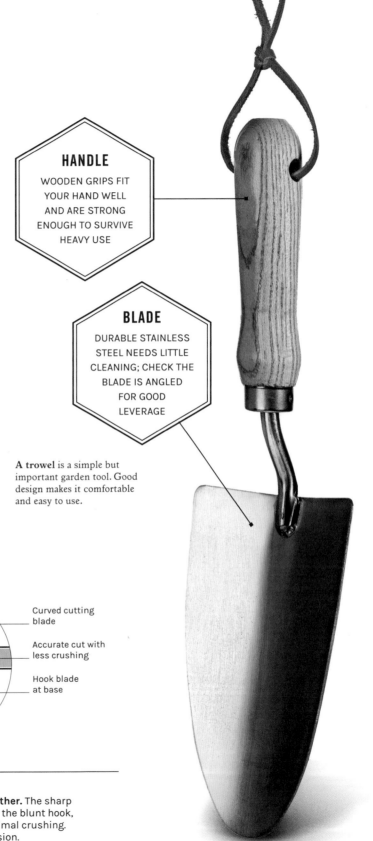

HANDLE
WOODEN GRIPS FIT
YOUR HAND WELL
AND ARE STRONG
ENOUGH TO SURVIVE
HEAVY USE

BLADE
DURABLE STAINLESS
STEEL NEEDS LITTLE
CLEANING; CHECK THE
BLADE IS ANGLED
FOR GOOD
LEVERAGE

A trowel is a simple but
important garden tool. Good
design makes it comfortable
and easy to use.

Stem

SIDE VIEW

Curved cutting
blade

Accurate cut with
less crushing

Hook blade
at base

BYPASS

Blades are curved and overlap one another. The sharp
edge of the top cutting blade slides past the blunt hook,
creating a smooth, clean cut, with minimal crushing.
Cuts can be made with precision.

FIRST SHOOTS

WHAT DO PLANTS NEED TO GROW?

It's easy to overcomplicate things when gardening. Like most living things, plants have very simple basic needs: light, air, water, food, and warmth. If your plants are wilting or brown at the edges, it is often simply because these needs aren't being met.

For every living thing larger than a microbe, oxygen is an essential, since it unlocks energy in food during a process called "respiration". You may have heard that animals inhale oxygen and exhale carbon dioxide, while plants do the reverse, taking in carbon dioxide and giving out oxygen. This is only half true. Plants also need to "breathe" oxygen, which they do through leaf pores called stomata (see p.30), giving out carbon dioxide, like us. Roots also need to absorb oxygen and will quickly suffocate in waterlogged or compacted soil.

SOLAR POWER

When the sun shines, plants make their own food (as sugar) in all their green-coloured parts, using a process called photosynthesis (meaning "light putting-together"). Zoom in and you will find green grains called chloroplasts (see pp.80–81) , which are stacks of light-harvesting machinery packed with green chlorophyll, where photosynthesis takes place.

Photosynthesis

Within green leaves, water is split in two inside light-absorbing stacks, releasing oxygen into the air and producing energy to power a mind-boggling series of reactions that culminates in carbon dioxide being fused together to make sugar.

WITHOUT WATER THERE IS NO LIFE

Water is not only needed for photosynthesis, but is also where the chemical reactions of life take place. Without liquid water, life grinds to a halt. Non-woody plants remain upright using water pressure (called "turgor") – each cell is like a taught, water-filled balloon, which deflates when water is scarce, causing wilting. It's the evaporation of water out of the pores in leaves that creates suction to draw the water and nutrients absorbed by roots up through stems to foliage and flowers. This flow is called transpiration (see p.30).

The more leaves a plant has, the faster water is lost, which is why large plants need more watering than seedlings. Water loss also goes up on warm and windy days when more moisture evaporates out of leaves, increasing the amount of water roots need from the soil (see pp.120–123).

KEEP THE TEMPERATURE JUST RIGHT

At its most basic level, life is merely a sequence of chemical reactions, much like music is a series of notes. The pace of these reactions is tied to temperature. Warmth speeds things up, but if it's too hot the proteins that drive life's reactions start to cook. Cold slows things down – photosynthesis pretty much stops below 10°C (50°F). Since plants can't make their own warmth or sweat to cool down, they rely on us to provide a suitable space to keep their notes playing life's melody.

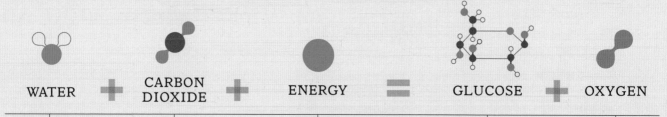

WATER	+	CARBON DIOXIDE	+	ENERGY	=	GLUCOSE	+	OXYGEN
Is transported from roots through stems into leaves.		**Absorbed from air** through stomata on the underside of a leaf.		**Sunlight is the energy** source that drives the photosynthesis reaction.		**The sugar** that fuels life processes in plant cells.		**A by-product** that is released into the air via stomata.

Respiration

Taking place within all cells, this vital reaction breaks down glucose to release the energy needed to power all of life's processes.

 **GLUCOSE** ➕ **OXYGEN** ═ **WATER** ➕ **CARBON DIOXIDE**

This energy source is created in leaves by photosynthesis.

Taken in from air through stomata on the underside of the leaf.

Evaporates out of stomata in leaves or is used for photosynthesis.

A by-product of respiration that can escape leaves via stomata.

SUNLIGHT
PLANTS USE THE SUN'S ENERGY TO POWER PHOTOSYNTHESIS AND MAKE SUGAR

CARBON DIOXIDE
ABOUT HALF THE CARBON DIOXIDE USED IN PHOTOSYTHESIS IS RELEASED BY RESPIRATION

WATER
IS DRAWN UP THE STEM AS IT EVAPORATES THROUGH STOMATA, IN A PROCESS CALLED TRANSPIRATION

OXYGEN
IS MADE IN LARGER AMOUNTS BY PHOTOSYNTHESIS THAN ARE USED IN RESPIRATION

Living, breathing leaf Each leaf is full of chloroplasts that make food from sunlight (see p.72), and pores that exchange gases with the air (see p.22).

HOW DO PLANT CELLS FUNCTION?

Zoom in on a leaf and you will see what looks like sheets of green-tinged bubble wrap, each bubble a living plant cell. Going closer still reveals each cell to be a bustling centre of life, made up of distinct parts with their own specific and extraordinary roles.

———

Individual cells are the tiny living units that all plants are made of and each of their microscopic components has an important role to play in keeping them, and the plant, alive and well.

CELL WALL

The tough, stretchy surround of each plant cell is made of fibrous cellulose – essentially a plastic made of sugar. Water flows through the wall's minute pores and embedded within it are sensors to detect stretch and movement (see pp.86–87), which trigger cell wall thickening and other chemical signals.

CELL MEMBRANE

Pushed against the inside of the cell wall, the cell membrane is an unimaginably thin layer of fat (lipid) that is vital for life. The molecular machinery studded across its surface constantly shuttles salts, sugars, and other substances in and out of the cell, as well as detecting hormones and potential threats.

CYTOPLASM (OR CYTOSOL)

Most of life's chemical reactions happen within this gloopy liquid that fills each cell. A scaffold of protein wires stretches throughout cytoplasm to the cell edges. This tethers the different organelles (mini-organs) in place, while various molecules glide like cable-cars along its length. Cytoplasm also contains sensors for light and heat.

VACUOLE

On its own, the cytoplasm inside a cell would not keep its cell wall taut enough to hold a plant upright. For this reason plant cells have developed vacuoles to give them extra rigidity. Like a balloon within a balloon, the membrane-lined chamber is filled with a pressurized, watery liquid called "cell sap", which holds the cell firm from within.

CHLOROPLAST

These minute, green, bean-shaped capsules harvest the sun's energy. Inside, pancake-like stacks of light-absorbing "thylakoids", crammed full of the green pigment chlorophyll, make food (a sugar called glucose) out of water and carbon dioxide, in a process called photosynthesis (see pp.70–71).

MITOCHONDRION

Smaller and simpler than chloroplasts, mitochondria are the biological power generators, which digest the sugar made from photosynthesis to supply energy for the cell. These organelles never rest and keep ticking over even at low temperatures and in dormant seeds.

NUCLEUS

This is the control centre which contains the cell's DNA, which has a protective double membrane of its own. All of the cell's activities are controlled by its molecular machinery, which reads the plant's genetic code within its DNA, and follows its instructions like a computer program. Chemical messages are ferried out of pores in the nucleus and all new substances made by a cell start life here.

THE CONSTRUCTION FACTORIES

A suite of structures – the endoplasmic reticulum, ribosomes, and Golgi apparatus – act together like a factory line to assemble new substances, following the genetic instructions sent from the nucleus.

Cell anatomy

An amazing array of structures makes up every
cell within a plant, all working together on
a microscopic scale to allow the plant to grow
and respond to its environment.

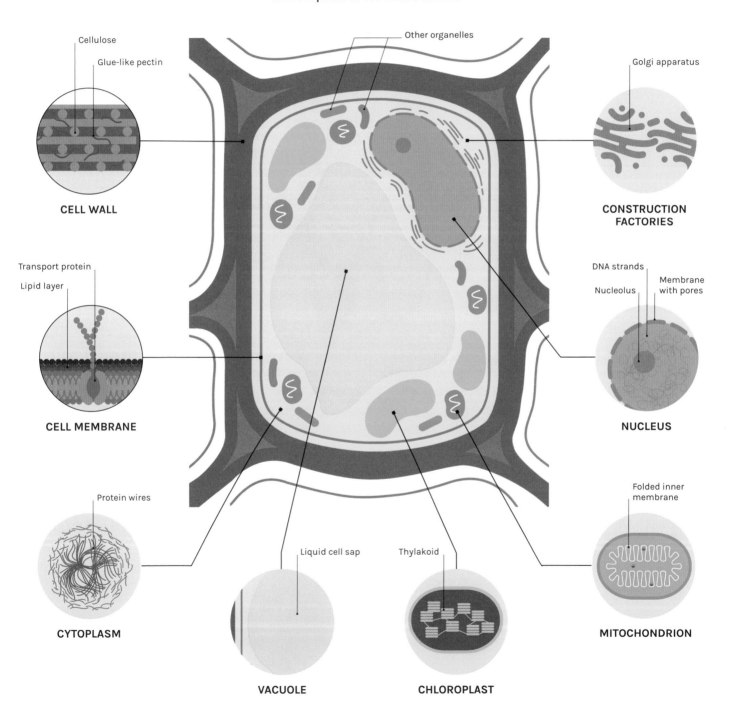

Cellulose

Glue-like pectin

CELL WALL

Transport protein

Lipid layer

CELL MEMBRANE

Protein wires

CYTOPLASM

Other organelles

Golgi apparatus

CONSTRUCTION FACTORIES

DNA strands

Nucleolus

Membrane with pores

NUCLEUS

Folded inner membrane

MITOCHONDRION

Liquid cell sap

VACUOLE

Thylakoid

CHLOROPLAST

WHAT IS A SEED?

Plants come in countless shapes and sizes, and so do their seeds, each tailored to the habitat it hails from. Yet every seed is fundamentally the same – a tiny, immature plant, called an embryo, encased in a protective shell.

———————

Before there were seeds, there were spores – tiny specks that spread through the air like a puff of smoke. Unprotected, these microscopic vessels of fragile genetic information give rise to the next generation only when they fall upon damp earth. Ferns and mosses use spores, but to spread beyond damp habitats, plants evolved seeds, which offer a protective coat for their precious cargo – a miniature plant, complete with its first one or two neatly folded seed leaves (cotyledons). In the mature seeds of some plants, like beans and peas, the cotyledons carry provisions to fuel early growth, while in many other plants, such as maize and onions, these stores are kept in a starchy bundle called the endosperm. The largest seeds, like the 20kg (44lb) coco de mer, take a huge investment of energy and are produced in small numbers, while orchids can create dust-like seeds by the thousand.

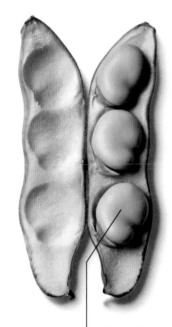

ADAPTED FOR SUCCESS

Germinating into this unforgiving world is always dicey and plants have evolved many strategies to tip the odds in favour of their seeds's survival. Unlike a human embryo, which will be born after nine months come what may, a plant embryo can lie dormant for months or years (even over 2,000 years), waiting for the perfect conditions for germination.

Plants have many methods to help seeds colonize fresh pastures far from parent plants. Many, like the winged seeds of maples (*Acer*) and parachute-like dandelion clocks (*Taraxacum officinale*), use wind-assisted flight. Seeds inside sweet fruit entice animals and birds to eat and deposit them after digestion, while hooked and barbed seeds snag on clothing or fur. Himalyan balsam (*Impatiens glandulifera*) seeds are fired 4m (13ft) from their pods and carried further in water.

A seed of two halves Broad bean seeds are made up of two cotyledons. This makes them easy to identify as part of the group of flowering plants called dicots (dicotyledons), the seedlings of which usually have two seed leaves. These are distinct from monocots (monocotyledons), which have a single cotyledon in their seeds and produce only one seed leaf.

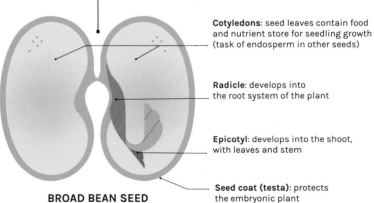

Cotyledons: seed leaves contain food and nutrient store for seedling growth (task of endosperm in other seeds)

Radicle: develops into the root system of the plant

Epicotyl: develops into the shoot, with leaves and stem

Seed coat (testa): protects the embryonic plant

BROAD BEAN SEED

HOW DEEP DO I SOW SEEDS?

New and seasoned gardeners may wonder whether they are planting their seeds at the optimal depth. Tradition says to cover a seed with its own depth of soil, and science reveals why this is a good rule of thumb.

Air and water need to percolate through the soil to stimulate seeds to germinate and for seedlings to survive. Small seeds may also need light. Planting too deeply, especially in dense, clay-rich soil (see pp.38–39) can suffocate a seed, as can overwatering, and pressing down (tamping) too hard on the soil after planting. Thin, sandy soils let air percolate more easily, and seeds may need to be planted a little deeper to receive the protection they need, while seeds in heavier clay soils may benefit from shallower sowing. Dry earth that has formed a hard crust can be too much of a barrier for a tiny seedling to punch through and should be loosened with a fork or hoe before sowing.

STARTING SMALL SEEDS

Scatter small seeds – 1mm (¹⁄₁₆in) diameter or less – on the soil surface or give them the lightest dusting of soil, because practically no light can penetrate through more than than 4–5mm (⅛–¼in) of heavy soil, or 8–10mm (⅜–½in) in light sandy soils. Given how easily light seeds can wash away in a downpour

or be pecked up by birds, small seeds are often best germinated indoors or in a greenhouse.

SOWING MEDIUM SEEDS

Medium seeds – 1–5mm (¹⁄₁₆–¼in) in diameter – do not usually need daylight and instead rely on temperature and water to trigger growth. Such seeds prefer a comfy blanket of soil or potting compost to offer protection from extreme temperatures, excess water, and peckish birds or rodents. Push medium seeds into the earth so they sit with at least their diameter of soil above them (2–10mm/¹⁄₁₆–½in) or by making a groove (also called a drill or furrow) and covering with the same depth of soil.

PLANTING LARGE SEEDS

Only when a growing shoot breaks the soil's surface can its baby green leaves – termed "cotyledons" – unfurl so that the seedling can start to feed itself by soaking in the sun's energy. Larger seeds of more than 5mm (¼in) diameter therefore carry enough food to sustain their shoots on a longer journey, allowing such seeds to be planted deeper and into safer surrounds. Planted at twice their length (about 5cm/2in deep for beans), large seeds can luxuriate in the moisture-rich soil, which is largely beyond the reach of smaller seeded plants' greedy roots.

Larger, soft-coated seeds such as beans and peas also benefit from this extra depth so they can drink in water slowly, since a deluge will cause them to swell rapidly, blistering and bursting their soft seed coat too quickly, and fatally gushing out their innards. Water seeps slowly through soil and so heavy rainfall will permeate slowly to deliver moisture gradually.

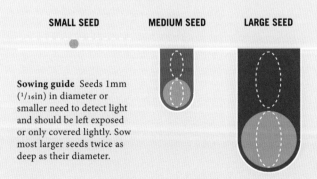

SMALL SEED   MEDIUM SEED   LARGE SEED

Sowing guide Seeds 1mm (¹⁄₁₆in) in diameter or smaller need to detect light and should be left exposed or only covered lightly. Sow most larger seeds twice as deep as their diameter.

SOWING DEPTH IS $2\times$ **THE SEED SIZE**

WHAT DO SEEDS NEED TO GERMINATE?

A seed may look dry and dead, but the germination process launches the tiny dormant embryo inside into growth so that a green shoot appears almost miraculously from where it was sown just days earlier.

Each seed has encoded in its DNA the conditions that are right for it to begin growth. It is the outer coat (testa) that senses its surroundings and holds the key to releasing this spring-loaded bundle of life. Germination cannot start until all the switches on the DNA checklist are flicked: for some it might be warmth after a period of cold that signals spring, or moisture after dry conditions. Water is absorbed through the testa, the seed swells, and the sleeping embryo starts growing – once the process of germination is set in motion there is no going back.

DRY SEED +

WATER + WARMTH + AIR (+ SOMETIMES LIGHT)

= GROWTH

CREATE CONDITIONS FOR LIFE TO THRIVE

Germination technically starts as soon as a seed starts soaking up water (imbibition) and ends the moment the baby root (radicle) pokes through the seed coat. Water, warmth, air, and sometimes light, are crucial for germination. When given these simple things many seeds will germinate straight from the plant or packet, although their specific requirements vary: some need heat, like *Amaranthus*, which germinates best at 35°C (95°F); while others, like kale, can germinate when soil is just 5°C (41°F). Sowing seeds indoors or in a greenhouse means you can shut the door on fickle weather and give seeds ideal germination conditions. A heat mat underneath a seed tray will supercharge germination,

speeding up the myriad chemical reactions. A cosy soil temperature of around 24°C (75°F) is optimum for germination in many garden plants. Covering an ordinary seed tray with a lid turns it into what's known as a propagator. This closed environment holds in warmth, increases humidity around seeds, and can contain an electric heater in its base, accelerating germination. After an initial watering, covered seed trays need little or no watering during germination. When a seedling has its first baby leaves (cotyledons) it can be taken off the heat to prevent it putting on rapid, soft growth that would be more prone to weather damage and pests.

BREAKING DORMANCY

Some seeds have extra safeguards to help ensure their survival, and gardeners can achieve more reliable germination by replicating the conditions particular plants experience in their natural habitats after falling to the ground. A simple method called "stratification", is often used to simulate the cold of winter or the warmth of summer for lengths of time using a fridge or a heated propagator.

Small seeds only contain a tiny cache of food and cannot afford to germinate when buried too deeply, because the seedling wouldn't make it to the surface. Many such seeds evolved to germinate only when

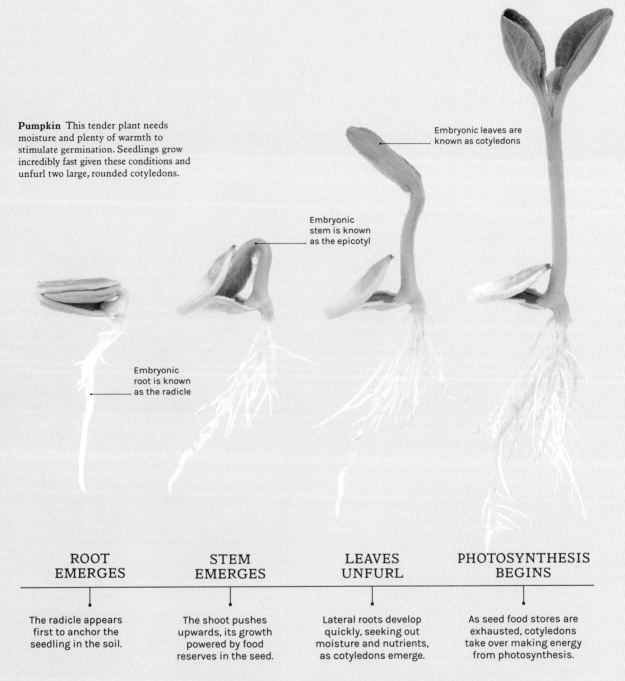

Pumpkin This tender plant needs moisture and plenty of warmth to stimulate germination. Seedlings grow incredibly fast given these conditions and unfurl two large, rounded cotyledons.

Embryonic leaves are known as cotyledons

Embryonic stem is known as the epicotyl

Embryonic root is known as the radicle

ROOT EMERGES	STEM EMERGES	LEAVES UNFURL	PHOTOSYNTHESIS BEGINS
The radicle appears first to anchor the seedling in the soil.	The shoot pushes upwards, its growth powered by food reserves in the seed.	Lateral roots develop quickly, seeking out moisture and nutrients, as cotyledons emerge.	As seed food stores are exhausted, cotyledons take over making energy from photosynthesis.

they can "see" the sun, via microscopic light sensors called phytochromes. These flick like a switch in sunlight, setting off a chain of chemical reactions to awaken the seed from its dormancy.

Seeds with a tough outer coat need some rough and tumble to damage the testa and allow the seed to imbibe water. In a process called "scarification", gardeners scratch, cut, or soak the surface of seeds like morning glory (*Ipomoea purpurea*) and sweet pea (*Lathyrus odoratus*) before sowing, to help break dormancy and initiate germination.

Almost unbelievably, some seeds will only germinate after a forest fire, when there is likely a fertile clearing for them to put down their roots. *Eucalyptus* and *Banksia* seeds are encased within cones or fruits sealed with resin, which melts in a fire's heat – a process that gardeners can mimic in an oven to extract seeds for sowing.

WHY DO WE SOW AND PLANT AT DIFFERENT TIMES OF YEAR?

Numerous factors influence the optimum times to sow and plant different types and species of plants, not least the quirks of your local climate. An understanding of these helps gardeners to get their plants off to the best possible start.

Over winter, the source of a plant's food – sunlight – is in short supply. Hardy plants that shed their leaves (deciduous) are "dormant" and don't grow over winter, and even those that retain their leaves (evergreen) grow very little because photosynthesis does not supply enough sugars when days are short. It is in this dormant state that plants will usually fare best when planted, lifted, or moved.

WHY PLANT DURING DORMANCY?

Most plants can survive being planted at any time of year, but planting in dormancy gives time for roots to establish so that they can efficiently absorb water and nutrients before spring growth puts serious demands on them. This holds true for deciduous trees, shrubs, and climbers, as well as plants that die right back in autumn and regrow in spring ("herbaceous perennials").

Late autumn is an ideal planting time, when soil is still warm and welcoming, while early spring also suits deciduous specimens. Woody plants can be planted through winter, as long as soil isn't frozen. Some nurseries take advantage of this dormant period to lift trees and shrubs (including roses) grown in fields, rather than pots, and ship them to customers with their soil-less roots exposed ("bare root") for immediate planting. This reduces the resources needed to grow and transport large plants.

WHY PLANT WHEN IN GROWTH?

Any plant growing in a pot can be planted when in growth, as it can be transferred without damaging its roots. However, because it is already in leaf, it must be kept well watered after planting, especially in dry, warm weather. Avoid digging up and transplanting established plants when they are growing, because damage to their roots reduces the water supply to the foliage, and plants may die (see pp.106–107). When planting in spring, check the last frost date in your area and ensure plants outdoors are hardy enough to cope if frost is still expected. Small bulbs, like snowdrops, dry out easily when dormant and are best planted in leaf, or "in-the-green".

HOW CAN I JUDGE WHEN TO SOW?

Sowing is almost always best done when soil is warm enough for growth, so germination begins quickly and seeds don't rot (see pp.82–83). For hardy plants, this can be as soon as the soil temperature starts to rise in early spring – the first flush of weed seedlings indicates that conditions are suitable. Vegetable crops can also be sown according to when you want to harvest: sow a few fast-growing radishes or salad leaves every few weeks through spring and summer for a succession of crops from mid-spring to autumn. Plants that flower in their second year (biennials), such as foxgloves (*Digitalis*), are often best sown as soon as their seed is produced in summer, so they have grown enough to flower the following year.

Tender plants should only be sown outdoors once the risk of frost has passed, but you can extend the growing season by sowing in a warm place indoors, or undercover in a greenhouse on a heat mat or in a heated propagator. Sow four weeks before the last frost date, after which young plants can be hardened off (see pp.86–87) and moved outdoors.

When to plant and sow

Cold temperatures slow down plants' internal processes, limiting growth and inhibiting seed germination (see pp.80–81), meaning it's often best to sow during the growing season and plant during dormancy.

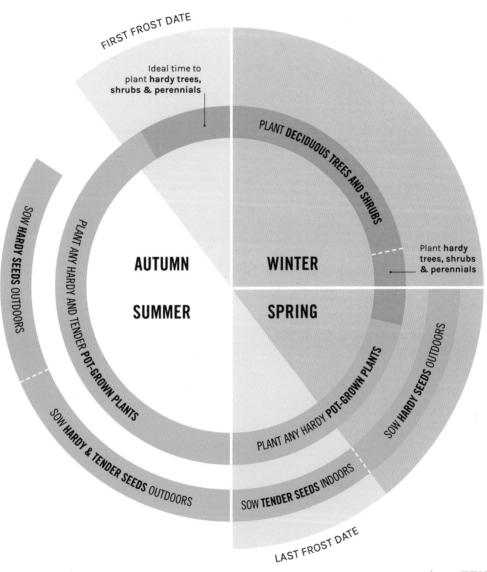

FIRST FROST DATE

Ideal time to plant **hardy trees, shrubs & perennials**

PLANT **DECIDUOUS TREES AND SHRUBS**

AUTUMN

WINTER

SUMMER

SPRING

Plant **hardy trees, shrubs & perennials**

SOW **HARDY SEEDS** OUTDOORS

PLANT ANY HARDY AND TENDER **POT-GROWN PLANTS**

SOW **HARDY SEEDS** OUTDOORS

SOW **HARDY & TENDER SEEDS** OUTDOORS

PLANT ANY HARDY **POT-GROWN PLANTS**

SOW **TENDER SEEDS** INDOORS

LAST FROST DATE

NO PLANT
GROWS BELOW

4°C
(39°F)

KEY

- Dormancy
- Growing season
- Frost period
- Frost risk period

FEW SEEDS
GERMINATE BELOW

7°C
(45°F)

WHAT IS MEANT BY "HARDINESS" AND HOW IS IT MEASURED?

Our ability to tolerate low temperatures is partly down to our genetics, and so it is with plant species – those that have evolved or been bred to cope in icy temperatures have this programmed into their DNA. How well a plant can withstand cold is termed its "hardiness".

Our love of growing exotic delights in temperate gardens means that many favourite plants aren't well equipped to handle cold weather and frost (see pp.118–119), but plenty of others are tough enough to stand long periods of freezing temperatures. The challenge for gardeners is to work out just how cold their garden is likely to get in winter and which plants will survive in these conditions.

HARDINESS ZONES AND RATINGS

Gardeners use general hardiness terms when talking about plants and these are a useful starting point: "hardy" or "fully hardy" plants will typically survive sub-zero winter weather; "half-hardy" specimens will succumb during a cold winter and may need some protection (see pp.160–161); "tender" plants have no tolerance for cold and will perish when winter comes. Of course, every region's winters and weather systems are different, and a range of hardiness maps and categories have been drawn up by experts to guide gardeners to plants that will flourish in their area.

Simple as this may sound, the reality can be confusing, as various organisations in different countries have created their own hardiness scales based on different criteria and naming systems. Since 1960, the United States Department of Agriculture (USDA) has published hardiness zone maps for the USA, dividing the country into 13 zones based on the average annual minimum winter temperature for each region. Gardeners therefore select plants assigned the same hardiness number as their zone or higher. These hardiness zones have been adopted in many countries. Canada has a similar system to the US, which also takes into account rainfall, frost-free days, and other variables.

In the UK, the Royal Horticultural Society (RHS) has created a detailed 9-rank system (H1a to H7, heated tropical greenhouse to very hardy), based on the lowest temperature that plants can survive. The European Garden Flora similarly rates plant hardiness, but into seven categories (H1 to H5, plus G1 and G2 for greenhouse plants).

LIMITATIONS OF CATEGORIZATION

None of these hardiness ratings or zones is foolproof, however, because it is difficult for them to factor in every variable. The length of time that plants can withstand low temperatures and the amount of winter rainfall can affect their survival. The growing conditions within a region may be quite different (see pp.34–37), so hardiness ratings can only ever be a general guide.

Neither do hardiness ratings take into account a plant's ability to handle summer heat and drought, and hardiness zones are not a reliable indicator of a region's climate for the rest of the year.

US – USDA ZONE RATINGS

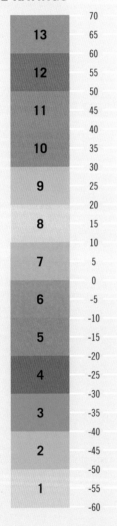

Zone	°F	°C
13	70	20
	65	
	60	15
12	55	
	50	10
11	45	
	40	5
10	35	
	30	0
9	25	
	20	-5
8	15	
	10	-10
7	5	
	0	-15
6	-5	
	-10	-20
5	-15	
	-20	-25
4	-25	
	-30	-30
3	-35	
	-40	-35
2	-45	
	-50	-40
1	-55	-45
	-60	-50

UK – RHS PLANT RATINGS

H1A Heated greenhouse – tropical

H1B Heated greenhouse – subtropical

H1C Heated greenhouse – warm temperate

H2 Tender – cool or frost-free greenhouse

H3 Half-hardy – unheated greenhouse/mild winter

H4 Hardy – average winter

H5 Hardy – cold winter

H6 Hardy – very cold winter

H7 Very hardy

EUROPEAN PLANT RATINGS

G2 Heated glasshouse even in south Europe

G1 Cool greenhouse even in south Europe

H5 Hardy in favourable areas

H4 Hardy in mild areas

H3 Hardy in cool areas

H2 Hardy almost everywhere

H1 Hardy everywhere

Hardiness ratings compared The RHS and European systems rate the hardiness of plants, allowing species and varieties to be categorized, and gardeners to judge what suits their climate. USDA zones are divisions based on minimum winter temperatures which are used as a guide to suitable garden plants in each area.

Acclimation

PLANTS CAN GRADUALLY ACCLIMATIZE TO THEIR LOCAL CLIMATE, adapting their inner workings to function better in cooler conditions. So plants raised locally are likely to be hardier than those imported from a warmer climate or grown in a heated polytunnel.

IS IT BEST TO SOW OUTDOORS OR UNDER COVER?

Deciding whether to sow outdoors, as nature intended, or "under cover" in a house, greenhouse, polytunnel, or other covered area, is important to give seeds the best start. Make your choice depending on the plant, the season, and the growing space available.

Sowing outdoors ("direct sowing") is strictly limited by the seasons, as seeds only germinate when the soil warms above 5°C (41°F) in spring, and seedlings may perish if caught by a late frost. Seeds always take longer to germinate and grow outdoors, when they risk being eaten by birds, mice, slugs, and snails

(see pp.202–203) and are vulnerable to unpredictable weather. But for many hardy vegetables and annual flowers, as well as tender plants in warm climates, sowing outdoors is quick and simple. It also doesn't require compost, containers, or undercover space. Carrots, parsnips, and other plants with taproots are

Planting year

Every seed has its own germination needs (see pp.76–77), and each garden is unique. During spring in a cool Northern hemisphere climate, soil temperatures are much higher under cover than outdoors, which benefits seed germination.

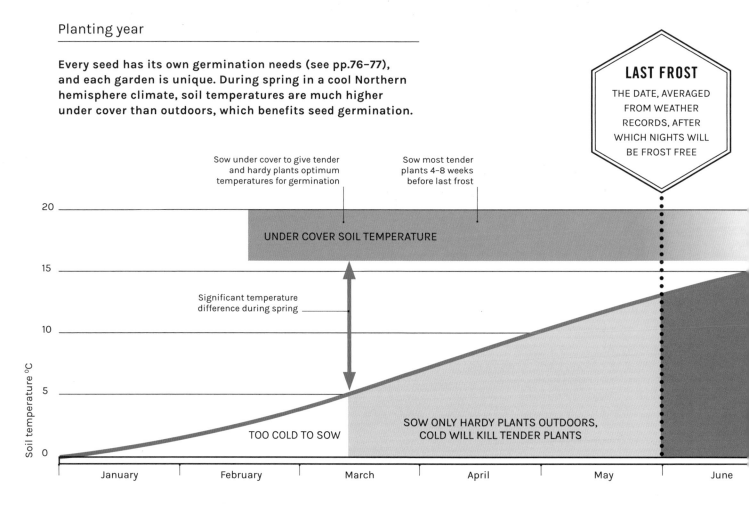

LAST FROST
THE DATE, AVERAGED FROM WEATHER RECORDS, AFTER WHICH NIGHTS WILL BE FROST FREE

Sow under cover to give tender and hardy plants optimum temperatures for germination

Sow most tender plants 4–8 weeks before last frost

UNDER COVER SOIL TEMPERATURE

Significant temperature difference during spring

TOO COLD TO SOW

SOW ONLY HARDY PLANTS OUTDOORS, COLD WILL KILL TENDER PLANTS

Soil temperature °C

20
15
10
5
0

January   February   March   April   May   June

always best sown directly, because their roots are easily damaged during "transplanting" (see p.85).

SHELTERED UNDER COVER

Starting seeds in seed trays, modules, or pots (see pp.84–85), filled with multipurpose compost or a special seed compost (see p.49) and kept under cover, means you can get a head start on growing while soil is too cold outside. This extends the growing season of all plants and is especially helpful to give tender annuals extra time to form flowers or ripen fruits in cooler climates.

Temperatures of at least 16°C (61°F) are essential to spur heat-loving tender plants, such as cucumbers and tomatoes, into growth, which means it's essential

to sow them under cover in areas where cold nights – or even frosts – can still strike in late spring. But even hardy plants typically germinate better at around 20°C (69°F), and you'll find that making spring sowings indoors on a warm windowsill, or in a propagator or on a heated mat in a greenhouse, gives faster, more reliable results than sowing outdoors.

This sheltered environment also keeps seedlings safe from pests, bad weather, and competition from weeds, boosting their chances of success. Seedlings grown under cover will, however, need to be hardened off (see pp.86–87) before being transplanted into their final positions outdoors (see p.85).

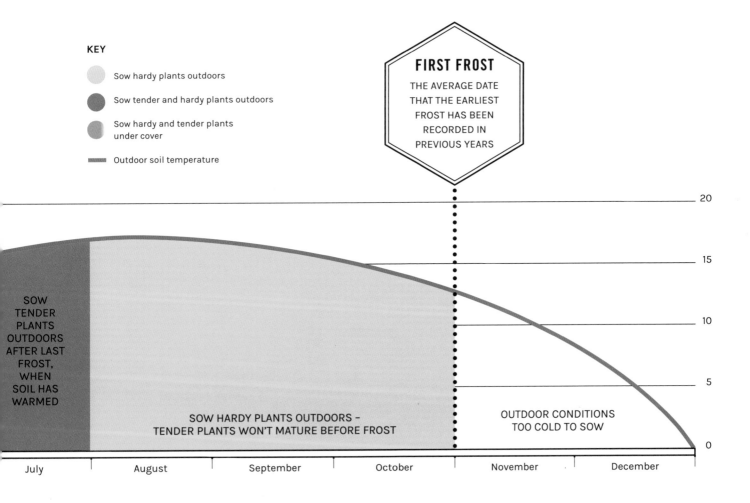

KEY

Sow hardy plants outdoors

Sow tender and hardy plants outdoors

Sow hardy and tender plants under cover

Outdoor soil temperature

FIRST FROST
THE AVERAGE DATE THAT THE EARLIEST FROST HAS BEEN RECORDED IN PREVIOUS YEARS

SOW TENDER PLANTS OUTDOORS AFTER LAST FROST, WHEN SOIL HAS WARMED

SOW HARDY PLANTS OUTDOORS – TENDER PLANTS WON'T MATURE BEFORE FROST

OUTDOOR CONDITIONS TOO COLD TO SOW

July | August | September | October | November | December

HOW CAN I KEEP MY SEEDLINGS STRONG?

Seedlings are delicate and have limited stores in their seed to fuel growth. Whether outdoors or under cover, they need your attention when young to make sure they have everything they need to grow big and strong.

The fight for survival starts early and seedlings growing close together will battle for water, nutrients, and light. Left alone, they will become tall and spindly ("etiolated") trying to outgrow their neighbours. There are different ways to avoid this, but the first is always to sow "thinly", leaving at least ½cm (¼in) between seeds, so seedlings are not cramped and are less likely to succumb to the fungal disease called "damping off".

THINNING OUT

Seedlings growing in rows outdoors or that have been sown thickly in pots or modules need "thinning out", which means ruthlessly culling them so that plants are spaced at the distance given on the seed packet. Many healthy seedlings may need to be sacrificed for the greater good, either by gently pulling them up or snipping the stem with scissors, leaving roots in the soil.

PRICKING OUT

Seedlings deliberately sown quite close together in seed trays or pots can be "pricked out" when very small and replanted in individual modules or pots. Use a small dibber (a stick with a rounded tip) or a pencil to ease roots from compost. Gently hold the seedling by a leaf, because damage to its stem would be fatal. Plant into multipurpose compost, which contains nutrients, rather than low-nutrient seed compost. Make planting holes with your dibber, gently firm compost around roots, and water well.

POTTING ON

Young plants quickly outgrow containers and need to be moved into larger pots (termed "potting on" or "potting up"), before their roots run out of space and struggle to draw in enough nutrients and water. This involves gently tipping the plant out of its pot and placing it in a larger one with fresh compost. How large this pot should be is your choice – science shows that the more freely roots can grow, the happier they are. Warnings that large pots cause roots to sit in water and rot are wholly wrong, unless the pot has inadequate drainage holes. Where space is limited, increase pot size gradually and pot up several times, but otherwise use larger pots that plants can grow to fill. Potting on also increases the space that leaves and stems have to grow, stopping them becoming etiolated.

TRANSPLANTING

Plants can be transplanted outdoors ("planted out") as soon as the weather and soil conditions are warm enough and they have been hardened off (see pp.86–87). Many seedlings thrive if transplanted just 4–6 weeks after sowing, but can also be potted on and planted out when larger. To transplant, dig a pot-sized hole in soil, tip the plant from the pot, place the roots in the hole, and gently firm soil around them. Water well to help roots establish. Seedlings often do well planted slightly deeper than in their pot, to help support their thin stems.

Choosing containers for sowing

Open seedtrays make good use of space, but seedlings will need to be pricked out into larger containers while small. Individual pots, and module trays require more room, but allow seedlings to be raised individually and transplanted with little root disturbance.

In containers

Many seedlings can be started off in a seed tray, but will need pricking out, and perhaps potting on, before transplanting.

Research shows that, on average, doubling pot size increases root and shoot production by 43 per cent.

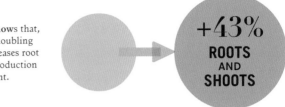

+43%
ROOTS
AND
SHOOTS

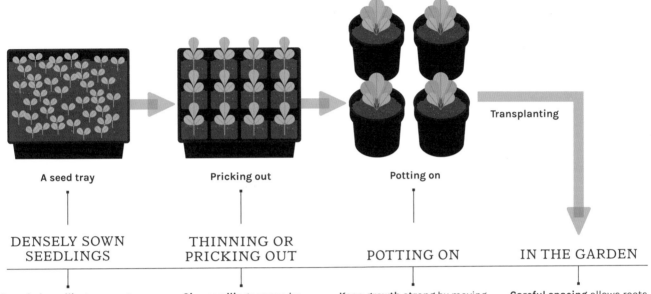

Transplanting

A seed tray

Pricking out

Potting on

DENSELY SOWN SEEDLINGS

THINNING OR PRICKING OUT

POTTING ON

IN THE GARDEN

Crowded seedlings compete for water, light, and nutrients and become weak or diseased.

Give seedlings space by removing unwanted plants or moving into modules.

Keep growth strong by moving young plants into larger pots with more water and nutrients.

Careful spacing allows roots to spread and strong growth resists pests and disease.

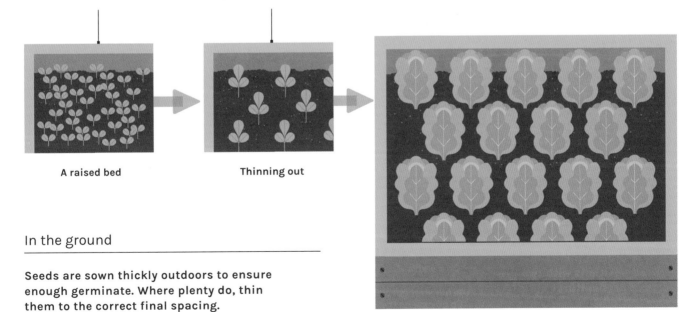

A raised bed

Thinning out

In the ground

Seeds are sown thickly outdoors to ensure enough germinate. Where plenty do, thin them to the correct final spacing.

Growing

WHAT IS "HARDENING OFF"?

A seedling or cutting raised under cover risks an untimely demise if moved outdoors without a bit of physical preparation beforehand. Acclimatizing plants to a life outside is called "hardening off".

It takes everyone a while to adapt to a different climate. After stepping off a plane, it can take weeks for our bodies to get used to unfamiliar heat, biting cold, or thin mountain air. A plant's internal biology is similar: a seedling cosseted in a cosy greenhouse or windowsill will be wholly unprepared for life outdoors.

SUN PROTECTION

Because window glass filters out the most powerful of the sun's rays (UVB), even seedlings grown on the brightest windowsills will not have experienced the full force of the sun (see pp.100–101). Like an untanned skin darkening in response to the sun, so leaves must produce sunlight-blocking chemicals, called flavonoids, through gradual exposure, before

they can tolerate full sun. Leaves lacking this protective shield in their outer layers will be pummelled by ultraviolet radiation, leading to sunburn ("sun scorch"). Sun scorched plants show browning of leaf margins or sometimes yellowing or darkening between leaf veins.

STANDING UP TO THE WEATHER

Mollycoddled under cover plants will similarly not have invested any resources into building layers of thick waxy protection or sturdy cellulose scaffolds to stabilize their stems. Cold and wind will come as a sudden shock, damaging or possibly killing the plant. Hardening off prepares plants for the unforgiving outdoors, gradually acclimatizing them to the physical assault

Hardening off

Plants from a windowsill or heated greenhouse should move first into a cold greenhouse or cold frame, if available. Alternatively, place them outside in a sheltered spot during the day and bring indoors at night, or protect plants in exposed positions with two layers of horticultural fleece initially.

UNDERCOVER

On a warm windowsill plants will not experience cold nights, a breeze, or even the full strength of the sun. Toughen them up by moving them outdoors gradually.

WEEK 1 – COLD FRAME PROPPED OPEN

Move plants into a cold frame with a glazed lid, which lets in light and can be propped open during the day to allow air movement. Close the lid at night to keep in warmth.

of moving out from under cover. While still indoors, prepare your green youngsters for exposure to wind by giving them a 10 second tickle once or twice a day. This stretches their cell walls enough to send a flurry of plant hormones through the plant, slowing upward growth and instead directing energies and nutrients into strengthening stems. The technical name for this is "thigmomorphogenesis" (see right) and commercial growers use fans or robots to achieve the same result. One plant hormone in this response is called salicylic acid – the active ingredient in aspirin. Research shows that a salicylic acid solution will trigger the same strengthening response when sprayed on leaves.

A GRADUAL PROCESS

There is no one-size-fits-all rule, but young plants can start to be taken outdoors in a step-wise fashion over two to three weeks, being careful not to expose tender plants to any frost that would kill them (see pp.118–119). Plants purchased from a nursery where they have been grown outside can be planted out immediately, although all others are best hardened off first.

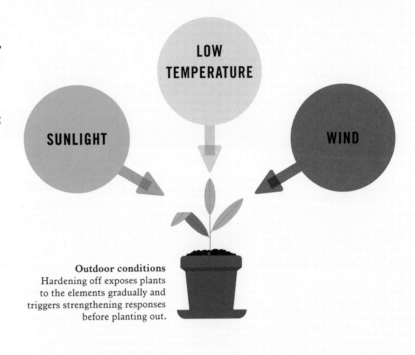

Outdoor conditions
Hardening off exposes plants to the elements gradually and triggers strengthening responses before planting out.

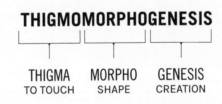

THIGMOMORPHOGENESIS

THIGMA TO TOUCH MORPHO SHAPE GENESIS CREATION

meaning "shaped by touch" is when plants change their growth pattern in response to touch, like that from rain drops.

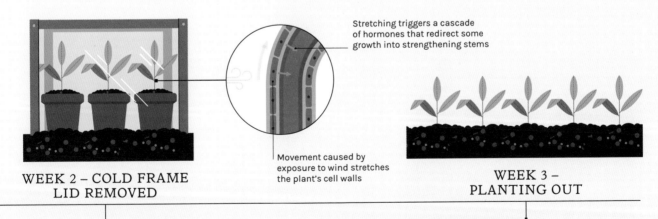

Stretching triggers a cascade of hormones that redirect some growth into strengthening stems

Movement caused by exposure to wind stretches the plant's cell walls

WEEK 2 – COLD FRAME LID REMOVED

As they become more accustomed to conditions in the cold frame, plants can be exposed to more light, cold, and wind by propping the lid wide open or removing it completely.

WEEK 3 – PLANTING OUT

Once hardened off your plants can be transplanted outdoors with much less likelihood of damage or cold shock. They may still benefit from fleece cover, however, if frost is forecast.

WHAT ARE F1 VARIETIES AND SHOULD I BUY THEM?

"F1 hybrid" or "F1" varieties produce identical plants with special features, such as pest resistance or large blooms. Some gardeners worry that this perfection comes from genetic engineering, but it is the result of labour-intensive breeding techniques.

In nature, individual plants from the same species reproduce with each other through the lottery of pollination (by insect or wind), to give offspring (termed "progeny") that are a 50:50 genetic mashup of mother (the seed bearer) and father (the pollen-giver). This means that if you collect and grow seeds from your garden, the resulting plants will all be different (see pp.176–177). Seed from such uncontrolled breeding is called "open-pollinated".

Growers use this natural method to produce the seed of non-F1 varieties, which include "heirloom" or "heritage" varieties that have been grown this way for generations (usually more than fifty years). To keep open-pollinated varieties pure, growers make

sure that no pollen from related varieties drifts in and that any plants with unusual features are removed ("rogued"). Some plants have flowers that can be fertilized by their own pollen ("self-pollination"), meaning that all offspring are usually very similar.

THE PRODUCTION OF F1 HYBRIDS

In contrast to open-pollination, plants can be deliberately "cross-pollinated" by humans taking pollen from one plant and brushing it on the female parts of another's flower. Features of mum and dad are passed on to the children: crossing red and white pelargoniums, for example, might give a mix of red,

Pollination

When pollen grains are transferred from male to female flower parts, pollination occurs. Depending on the species, this may take place on a single plant (self-pollination), or between different plants (cross-pollination), both of which can produce healthy seeds.

INDIVIDUAL SELF-POLLINATION

In some bisexual flowers successful pollination can occur within individual blooms.

SINGLE PLANT SELF-POLLINATION

Pollen transferred on a single plant gives little genetic variabilty and offspring that are alike.

CROSS-POLLINATION

Pollination between plants, combines the genes of both parents in hybrid offspring that may vary.

Creating an F1 hybrid

Producing F1 seeds is a long and meticulous process that culminates in the cross-pollination of two parent lines to create the first filial (F1) generation.

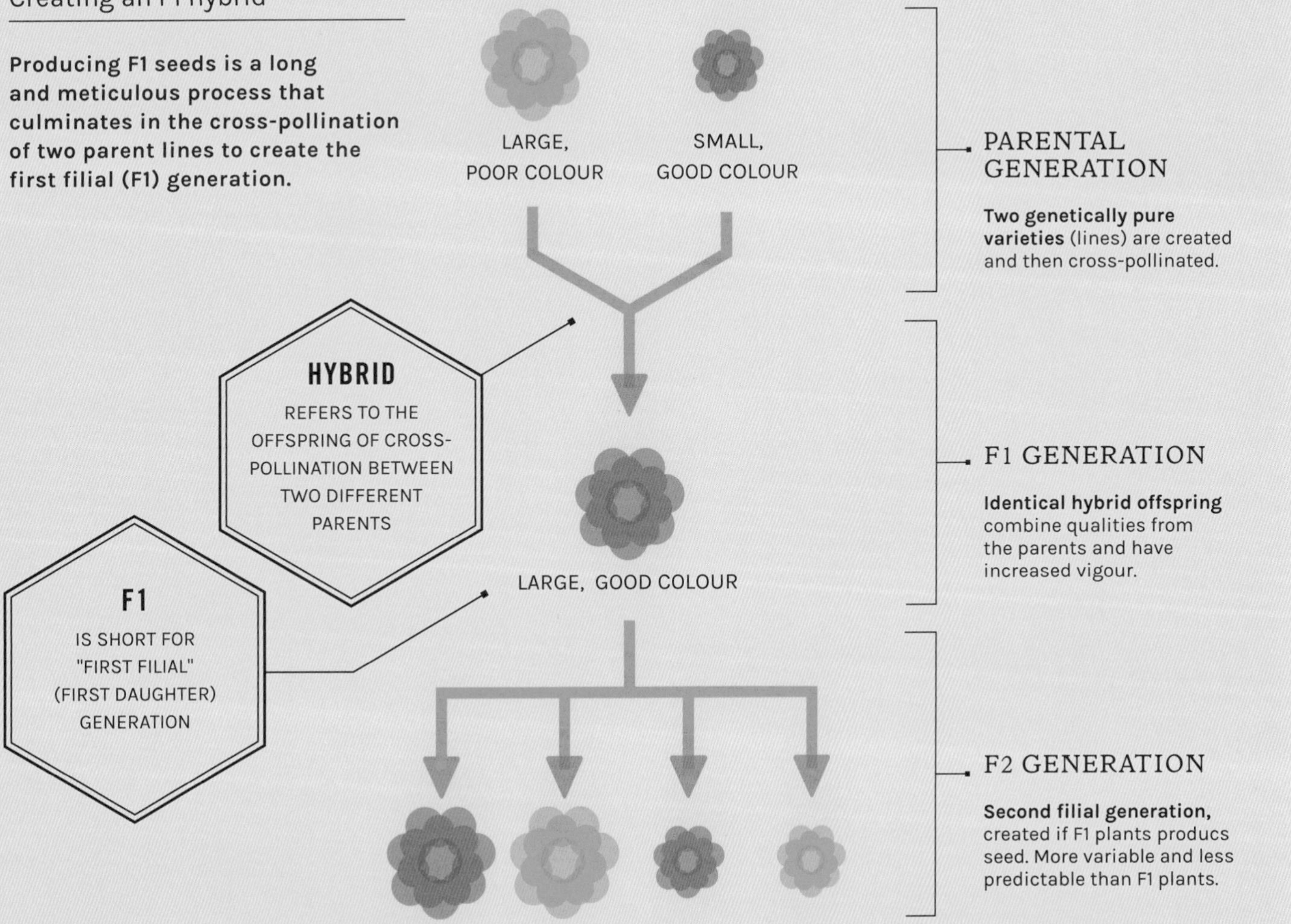

LARGE,
POOR COLOUR

SMALL,
GOOD COLOUR

**PARENTAL
GENERATION**

Two genetically pure varieties (lines) are created and then cross-pollinated.

HYBRID

REFERS TO THE OFFSPRING OF CROSS-POLLINATION BETWEEN TWO DIFFERENT PARENTS

F1

IS SHORT FOR "FIRST FILIAL" (FIRST DAUGHTER) GENERATION

LARGE, GOOD COLOUR

F1 GENERATION

Identical hybrid offspring combine qualities from the parents and have increased vigour.

F2 GENERATION

Second filial generation, created if F1 plants producs seed. More variable and less predictable than F1 plants.

white, and pink offspring. The value of F1 hybrids is that they have none of this variation – what grows is exactly what's shown on the seed packet.

To do this, breeders painstakingly create two genetically pure parent varieties that they then cross-pollinate to produce seed. They might start with a marigold that has a particularly large, but poor-coloured flower, and another with a rich orange colour but a small flower. These plants are self-pollinated to create two "inbred lines", where parent and child are almost identical, which may take eight years. The two inbred lines are cross-pollinated by

hand to create F1 seed, which will give large, orange marigolds, with very little variation between plants. The offspring of two inbred parents can also be stronger (called "hybrid vigour").

All this time and care make F1 hybrids expensive, but can give them improved disease resistance, size, colour, and yield. Having identical plants can also have its drawbacks: F1 varieties of some vegetables, such as cabbages, tend to mature simultaneously, leading to a glut. Also, don't bother trying to collect seeds from F1 hybrids, as their offspring will be highly variable and often inferior to the parent plant.

WHAT ARE GRAFTED PLANTS AND SHOULD I BUY THEM?

Stitching someone else's limb onto your body sounds like science fiction, but has been practised with plants for over 2,500 years. Termed "grafting", fusing two different plants is widely used to control size, disease resistance, and vigour.

Many trees, shrubs, roses, and some vegetable plants are often grafted, but because only closely related plants can be grafted together, there are no horses' heads on lions' bodies! Members of the monocot group of plants (including bamboos and palms) are almost impossible to graft.

HOW IS GRAFTING DONE?

This technique involves carefully lining up the cut surface of a plant shoot ("scion") with the decapitated stem of a rooted plant ("rootstock"). Grafting is possible thanks to

a dense green ring of stem cells (see pp.136–137) under the bark, called the cambium, which runs alongside the plant's sap-carrying piping (called xylem and phloem). Grafters use a sharp knife to make matching angled cuts on the rootstock and scion, and bind them together with tape or wax.

THE ADVANTAGES OF GRAFTING

As scion and rootstock grow together, the scion stays a genetically separate plant, producing its own leaves, flowers, and fruits. The rootstock supplies the scion with water and nutrients, hormones that control

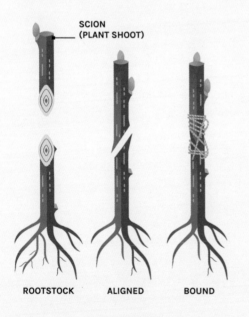

SCION
(PLANT SHOOT)

ROOTSTOCK   ALIGNED   BOUND

Uniting rootstock and scion
The two surfaces must be cut at matching angles so they fit together neatly. The graft is then bound to hold it in place.

Grafting

After a graft is bound, the two rings of stem cells (cambium) from rootstock and scion (plant shoot) will form callus cells, which bridge the gap to unite them. This allows water and sap-carrying tissues to join, and the plant to grow.

KEY

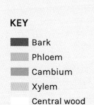

■ Bark
▨ Phloem
▨ Cambium
▨ Xylem
□ Central wood

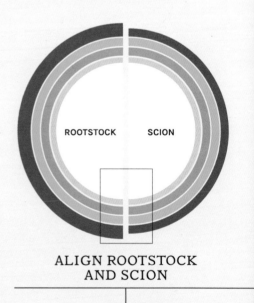

ROOTSTOCK   SCION

ALIGN ROOTSTOCK AND SCION

Careful positioning of the two sections is needed so that the rings of cambium cells that will unite them line up.

growth, and can also improve its resistance to disease and drought. Fruit growers have taken advantage of this by breeding rootstocks that control the final size of fruit trees. Rootstocks for apples range from the "very dwarfing" M27, which produces small, weak-growing trees that need good soil and permanent staking (see pp.130–131), to the "very vigorous" M25, which forms trees too large for most gardens.

Like cuttings, grafting is a type of "vegetative propagation" that allows dozens of plants to be made from one shrub or tree (see pp.136–137). But, unlike cuttings, grafted plants have a pre-formed root system and so will develop faster. The skill and time needed makes grafted plants more expensive, but this investment is worthwhile for a long-lived tree or shrub. Cucumbers, tomatoes and other vegetables can be grafted onto rootstocks chosen for their speed of growth or disease resistance. Only consider paying the hefty premium for these if you have experienced problems with conventionally grown plants.

Rootstock suckers

WATCH OUT FOR NEW SHOOTS CALLED "SUCKERS" GROWING FROM THE ROOTSTOCK OF WOODY PLANTS (SEE PP.174–175).
These vigorous growths from below the graft have leaves and flowers that differ from those of your chosen variety, which come from the scion. Suckers can quickly come to dominate a plant. Rub them out or pull them off when you see them.

WHAT TO LOOK FOR

You can usually spot a graft as a ring-shaped or angled scar in the bark, or a slight difference in bark colour or texture between rootstock and scion. With roses, the graft may be below soil level. Some weeping trees are grafted at a height of 1.5–2m (5–6ft), to give branches space to droop.

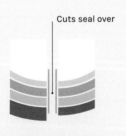

Cuts seal over

REPAIR BEGINS

Before a graft union forms the healing response seals the wounds to prevent sap loss and infection.

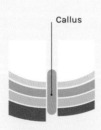

Callus

CALLUS BRIDGE

Both scion and rootstock produce cells to create a scar called a callus to bridge the gap.

Cambium bridges graft

CAMBIUMS JOIN

Cells from the cambium divide and unite across the callus bridge and can then form other cell types.

Phloem and xylem join

GRAFT COMPLETE

Xylem and phloem connect so that water and nutrients can flow through the stem.

HOW MUCH SPACE SHOULD
I LEAVE BETWEEN PLANTS?

Being squashed into a busy metro train is bad enough, but for an immovable plant, being crammed in close to a neighbour can make the difference between barely surviving and thriving. Spacing is something that gardeners need to judge carefully.

MYTH
- TREE ROOTS MIRROR THEIR CANOPY

TREE ROOTS CAN SPREAD

4–5 X

AS WIDE AS THE LEAFY CANOPY

The size and strength of a plant is wedded to the spread of its root system. Roots are a plant's sole source of water and soil nutrients, and if they are confined a plant's growth will be stunted – seen most dramatically in bonsai trees. Where roots have freedom to spread, plant growth is limited only by the nutrients and water within the soil.

Any nearby root system will compete for water and nutrients, and will restrict growth if competition is too fierce. Adequate plant spacing is therefore essential if plants are to reach their full potential. This relies upon knowing how wide their roots, as well as their foliage, will grow.

HOW BIG WILL EACH PLANT GET?

In a garden, trees are the greediest plants, and therefore ideally need a wide berth from other trees and shrubs. Only plants that can cope with dry, shady conditions should be planted underneath them, especially as tree roots are mostly in the top 30cm (12in) of soil, and so tend to suck it dry (see left).

Pay attention to the final size of plants when they are mature, which is often stated on plant labels. Space long-lived trees and shrubs this distance apart so that they can grow without the need for pruning to restrict their size. Woody climbers also benefit from generous spacing so that

their roots can spread and support strong growth. Plants that can often be planted closer together are annuals, which have only one growing season to spread their roots, and many herbaceous perennials, which can be lifted and divided (see pp.182–183) when they begin to outgrow their space. Denser planting also helps cover the soil and smother weeds.

SPACE TO SUIT YOUR NEEDS

There are no absolute rules when it comes to plant spacing – closer planting will simply mean smaller plants. Moreover, the quality of the soil holds great sway. If soil is kept healthy with regular mulching (see pp.42–43) then "multisowing" some vegetables, such as beetroot, radishes, and onions, can reap fantastic results. Sowing three of four seeds together and planting them out as one, will yield a cluster of slightly smaller, but perfectly formed plants, and allow you to plant at a higher density than you can with single plants.

Raised beds make efficient use of space, because they can be packed with crops without having to leave space to walk on soil. You can fit more plants in by arranging rows along a plot's length, rather than across its width. Grid planting patterns make the most efficient use of space, especially when plants are arranged in triangles rather than squares (see right).

Planting patterns

Arranging plants carefully gives them room to grow and thrive, and allows you to fit more in and maximize the potential of your space. This is especially useful for vegetables, but works in flower beds too.

KEY

● Size at planting

○ Final size

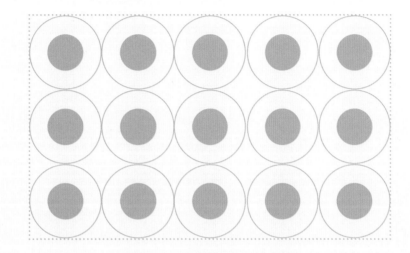

SQUARE GRID

Aligning neighbouring rows of plants makes it simple to measure the right planting distance, but unused space is left between rows.

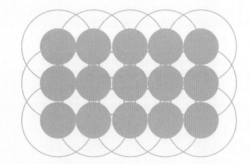

CONGESTED

Planting too close together leaves little space for growth. For any plants to do well, some will need to be removed.

TRIANGULAR PATTERN

Staggering rows in a triangle grid is efficient and allows 22 per cent more plants to be planted in an area than a square grid.

ARE SOME PLANTS BETTER GROWN TOGETHER?

Growing plants together for their shared benefit is known as "companion planting". While the practice has long been a melting pot for folklore and wishful thinking, scientists have found that some plants really are able to help others.

Plants from the pea and bean family ("legumes") can take nitrogen from the air, thanks to a beneficial relationship with soil bacteria living in nodules on their roots (see p.122). Some of this nitrogen leaks into soil to nourish nearby microorganisms and plants, and will be bequeathed to the earth when they die and rot down. Research shows that potatoes grow significantly larger alongside French beans, as does lettuce grown near peas. Legume cover crops (or "green manures") can be sown between plantings to add nutrients to soil and crowd out weeds.

ATTRACT AND REPEL

Companion plants can be grown to lure, trap, confuse, and repel would-be insect pests. For example, dill, fennel, and other plants in the carrot family (Apiaceae), have heads of many tiny flowers that attract aphids, as well as their predators, including hoverflies. When used to edge a plot, these companions pull pests away from plants like a trap. The strong scents of marigolds, sage, and other herbs can mask the smell of prized plants alongside them, confusing or even repelling pests. Scientists in the UK and Africa have developed combinations of companion plants that create effective "push-pull" systems to reduce pest damage on commercial crops (see below), so it may be worth exploring what works in your garden. Clever plant pairings can also entice pollinators to boost fruit crops. For example, attract bumblebees with "hooded" snapdragon flowers.

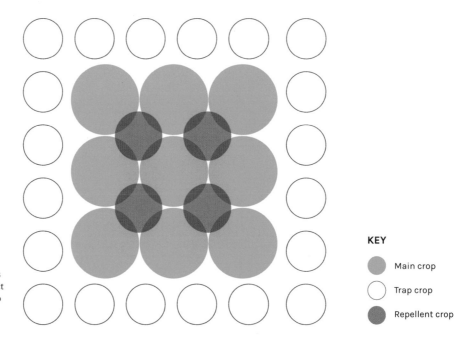

"Push-pull" systems protect crops using perimeter trap crops to attract pests and strongly scented plants to repel them, increasing some maize yields in Africa by 80 per cent.

KEY

- Main crop
- Trap crop
- Repellent crop

SHOULD I ROTATE PLANTS EACH YEAR?

The value of cycling where crops are grown in a "rotation" has been known for 3,000 years, but many modern guides on this practice might not be well rooted in science. It's likely, however, that a simple rotation may benefit the health of your plants.

Rotation based on nutrient needs
Alternating crops that are hungry for specific nutrients helps prevent soil becoming depleted. Use compost mulch to replenish soil nutrients rather than nitrogen-fixing crops.

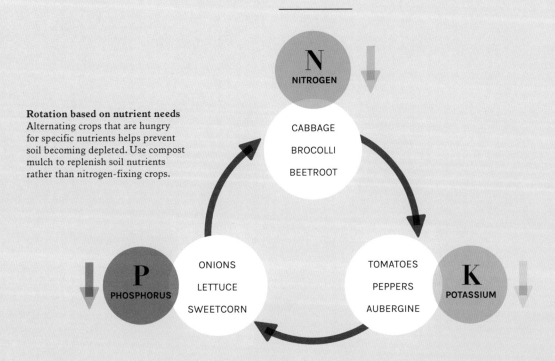

N
NITROGEN

CABBAGE
BROCOLLI
BEETROOT

P
PHOSPHORUS

ONIONS
LETTUCE
SWEETCORN

TOMATOES
PEPPERS
AUBERGINE

K
POTASSIUM

All plants plunder soil for the nutrients they need to grow, but each has a different diet. For example, all except legumes (see p.94) drain the soil of nitrogen, potatoes are hungry for potassium, while lettuce uses lots of phosphorus to help provide energy for leaf growth. It therefore makes sense to change where crops are grown each year to vary what is taken from soil and avoid nutrient deficiencies. Today, many farmers use a science-based cycle that alternates cereal crops with a nitrogen-fixing green manure, like alfalfa (aka lucerne or *Medicago sativa*), which is ploughed back into soil. This replenishes soil nitrogen, but in gardens where growing space is at a premium, soil may not remain empty for long enough to grow green manures (see pp.42–43).

AVOID PEST AND DISEASE PROBLEMS

Another reason to rotate crops is to reduce problems with the handful of troublesome pests and diseases, such as clubroot, onion white rot, and potato cyst nematodes, that are able to lie dormant in soil or plant litter, attacking again the following year – or even several years later – if another susceptible crop is planted. Plants from the same family are often prone to the same problems, which is why it's often recommended to rotate related plants together as a group. It's not just vegetables though: plants from the rose family (Rosaceae) are also afflicted by "replant disease", when planted after one another. The cause of this stunted growth is unknown, and the only solution is to plant unrelated species.

WHAT'S THE DIFFERENCE BETWEEN A BULB, CORM, TUBER, AND RHIZOME?

Bulbs are just one of several different types of food storage organ that plants have evolved as subterranean fuel tanks. These are packed with energy to enable plants to survive dormant periods and power rapid growth and flowering when conditions are most favourable.

ROOTS
NEW ROOTS GROW FROM EXTENDING RHIZOMES, SUPPLYING EACH SECTION WITH WATER AND NUTRIENTS

Any plant that keeps its store of fuel reserves underground and sprouts on or beneath the earth's surface is technically called a "geophyte" (meaning earth-plant). Plants with bulbs are geophytes, but many of what most gardeners call "bulbs" are not actually bulbs at all. Earth-plants are categorized into four groups – bulbs, corms, tubers, and rhizomes. Originally evolving from leaves or stems, each is an underground stockpile of stored sugar produced during photosynthesis (see pp.70–71). The underground parts of these plants also multiply and spread, and are often easy for gardeners to divide into new plants (see p.187).

RHIZOME
FLESHY STEMS OF BEARDED IRISES SPREAD ALONG THE SOIL SURFACE, SENDING UP SHOOTS FROM NEW SECTIONS

WHEN IS A BULB NOT A BULB?
Although they may look similar, the internal structures and growth of geophytes are quite different. Learning to distinguish between them will help you to plant and propagate them correctly.

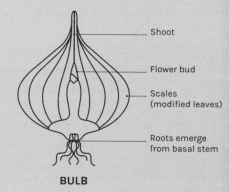

Shoot

Flower bud

Scales (modified leaves)

Roots emerge from basal stem

BULB

BULBS

If you've ever cut into an onion, then you'll know that a bulb has a papery coat covering layers of soft flesh. Each layer is actually a pale leaf (called a "scale"), drained of its sun-harnessing greenness and bulging with stored energy. Beneath the soil, bulbs lie in wait until the time is right for the flower bud at their centre to sprout upwards and roots to emerge from the flat base. Fuelled by the energy in its scales, the shoot unfurls into a bright bloom and lush leaves, which soak up sunlight like solar panels to replenish the bulb's energy store ready for the next growing season.

CORMS

Outwardly corms look like true bulbs, with one central sprouting point and fleshy roots emerging from a flat base. Slice through a corm, however, and you will find solid, energy-dense flesh, made from a stem base that has swollen into an underground sack of starchy fuel. As corms are planted and cared for like true bulbs, gardeners and retailers often group them together.

TUBERS

Potatoes, some *Anemone* species, and *Caladium* are examples of "stem tubers", where the solid flesh of the storage organ is formed from an underground stem that has inflated, sometimes to huge proportions. A stem tuber can sprout at various points across its surface, forming little buds called "eyes". Sprouting will initiate from the end with the most eyes (the "rose end").

Other geophytes, including sweet potato (*Ipomoea batatas*) and dahlias, have enlarged "tuberous roots" as their energy source. These form buds that grow into new shoots at the end where they were attached to the old plant. Although root vegetables, such as carrots, also have a thick root (called a tap root), they aren't called tubers since they are not perennial and don't divide underground.

RHIZOMES

Horizontal underground rhizomes of ginger (*Zingiber officinale*), canna, and bamboos are ancient geophytes – having evolved before dinosaurs roamed the earth – and are better able to cope with extreme temperatures and drought than other "bulbs". What look like gnarly roots are actually thick stems, which spread sideways and store energy. They can sprout along their length, sending shoots upwards to form what looks to all appearances like a new plant (termed a "sucker") on the surface, but these remain attached to the mother plant.

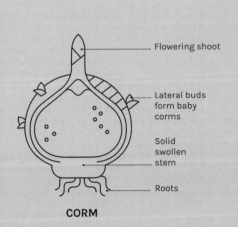

Flowering shoot

Lateral buds form baby corms

Solid swollen stem

Roots

CORM

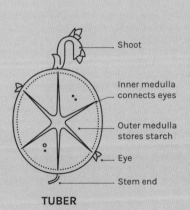

Shoot

Inner medulla connects eyes

Outer medulla stores starch

Eye

Stem end

TUBER

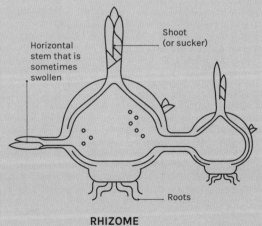

Horizontal stem that is sometimes swollen

Shoot (or sucker)

Roots

RHIZOME

HOW DEEP SHOULD
I BURY A BULB?

Every gardening book or website will advise you on the "correct" depth to
plant bulbs, but don't fret about needing a tape measure because – unbelievably –
many have their own ways of reaching their preferred soil depth.

Generally speaking, most bulbs, corms, tubers, and
rhizomes are best planted in a hole two to three
times deeper than than they are tall. There are
always exceptions, however: scientific research shows
that you may get better blooms from tulips planted
up to 20cm (8in) deep; and the rhizomes of bearded
irises (*Iris germanica*) are best planted shallowly so
that their tops show above the soil and are exposed
to sunlight, which improves flowering.

Back-of-the-packet instructions will likely give you a
good outcome, but precision is not essential because
bulbs have "contractile roots", which have an almost
magical ability to stretch and contract like a muscle,
pulling the bulb into the earth to its preferred depth.
Over several weeks they will ease themselves into the
damp safety of the soil, out of sight of rodents like
mice and squirrels, which often view energy-rich
bulbs as an easy meal.

WHICH WAY UP?

Many bulbs have a pointed top where the tip of
the flower bud emerges, and a woody, flat bottom
(termed the "basal plate"), out of which roots grow.
When this is obvious it makes sense to place them in
the ground this way up, but when these features are
unclear there is no need to worry about making sure
you plant a bulb the "right way" up. Just as with
seeds, you could plant most bulbs upside down and
they would be absolutely fine – roots follow gravity
and always grow down, and shoots can sense which
way is up. Try it yourself: plant half of your daffodil
bulbs pointy end down and the other half the right
way up in autumn. When they flower in spring you
are unlikely to be able to see the difference – upside
down bulbs may be slightly shorter. Using their
contractile roots bulbs will also shift themselves
upright over many months, so that the growing
point is eventually uppermost. Many tubers and
rhizomes, especially those planted quite shallowly,
are not able to right themselves and will only
flourish when planted in the correct orientation.

WINTER ACONITE   **CROCUS**   **DAFFODIL**   **LARGE ALLIUM**

10cm (4in)

20cm (8in)

PLANTING DEPTH 3X THE BULB HEIGHT

Bulb planting guide Gardeners tend to plant
bulbs too shallow, especially if soil is hard or
stony. This can result in fewer flowers, so plant
at three times the bulb's height where possible.

30cm (12in)

WHAT CAUSES A BULB TO SEND OUT SHOOTS?

Within each of a bulb's cells a biological clock uses cues from above ground to track the passing seasons, enabling bulbs to time flowering to maximize the chance of reproductive success.

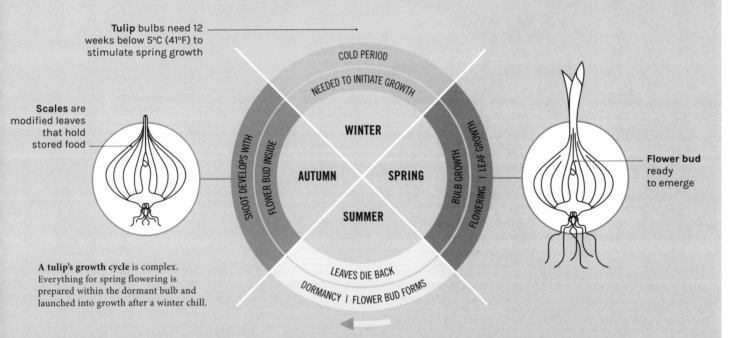

Tulip bulbs need 12 weeks below 5°C (41°F) to stimulate spring growth

Scales are modified leaves that hold stored food

A tulip's growth cycle is complex. Everything for spring flowering is prepared within the dormant bulb and launched into growth after a winter chill.

COLD PERIOD
NEEDED TO INITIATE GROWTH

WINTER

AUTUMN — SPRING

SUMMER

SHOOT DEVELOPS WITH FLOWER BUD INSIDE

BULB GROWTH

FLOWERING | LEAF GROWTH

LEAVES DIE BACK
DORMANCY | FLOWER BUD FORMS

Flower bud ready to emerge

A bulb's complex biological clock sends an array of chemical signals through its tissues in response to soil temperature and any light that reaches it from the surface. Each plant has evolved a pattern of growth and rest that reflects the climate and growing conditions of its native habitat. Spring-flowering bulbs, such as crocuses and tulips, typically evolved in habitats with harsh, dry summers and so grow and reproduce rapidly in spring, before retreating underground to wait out the summer. They then use winter as the cue to restart the cycle the following year.

VERNALISATION

Plants have sophisticated ways of tracking the changing seasons and many rely on exposure to a long cold spell during winter to trigger flowering – a process known as

"vernalisation". Gardeners have long known that bulbs hailing from temperate regions need to experience a period of low temperatures to flower, but the processes behind this remain poorly understood. Despite apparently being dormant in autumn and winter, many bulbs make use of their energy reserves to start growing shoots and flower buds during this period.

Understanding vernalisation means that you can trick those that need a cold period into flowering in regions with warm winters, or out of season, just by chilling them in a fridge for eight to fifteen weeks. When planted in moist soil or even a water-filled jar, bulbs "forced" in this way will grow and flower in mid-winter if you choose. Not every bulb needs chilling, amaryllis (*Hippeastrum*), for example, will flower whenever it is moist and warm.

HOW CAN I WORK OUT THE BEST SPOT FOR GROWING PLANTS INDOORS?

No plant evolved for life inside a modern building, but knowing a houseplant's wild origins will help you select a spot that's a home from home. While some plants are adaptable, others won't thrive without the perfect location.

———————

Houseplants hail from all over the globe and when deciding where to place them, consider whether the humidity, light, and temperature matches their natural habitat. Many are native to subtropical forests, where temperatures and humidity are high but light levels vary from very low on the forest floor, where *Fittonia* and *Anthurium* flourish, to much brighter in the canopy where orchids and many bromeliads (pineapple family plants) cling to branches. Plants from arid habitats, such as cacti and succulents, love low humidity, hot sun, and seasonal rainfall. Mediterranean-type climates are home to plants including *Strelitzia* and *Hippeastrum*, which need sunny conditions, with temperature and moisture levels that fluctuate seasonally. Pot azaleas and a few other houseplants come from temperate regions and prefer slightly lower light levels, in relatively cool, moist conditions.

Take into account how large a plant will grow and whether it's a climbing plant that will need support, or has tumbling stems that would suit a position on a shelf or in a raised planter.

GET LIGHT RIGHT

We get calories from food, but plants get theirs when their leaves capture "photons" of sunlight, some of which are filtered out even by the cleanest, clearest window glass (see right). The intensity and duration of natural light varies with the seasons: in autumn and winter many plants need to be moved close to windows to give them enough light, but in the

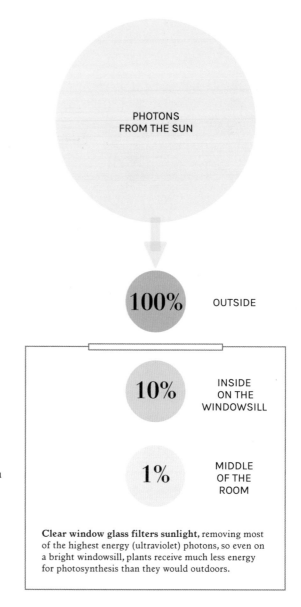

PHOTONS FROM THE SUN

100% OUTSIDE

10% INSIDE ON THE WINDOWSILL

1% MIDDLE OF THE ROOM

Clear window glass filters sunlight, removing most of the highest energy (ultraviolet) photons, so even on a bright windowsill, plants receive much less energy for photosynthesis than they would outdoors.

longer, brighter days of spring and summer, many are best moved out of intense sunlight. North-facing windows also receive less light than those facing south.

Most houseplants fare well on the sill of a well-lit window or on a table, shelf, or stand about 50cm (20in) away from a bright, south-facing window. Some direct sun suits succulents and many plants from Mediterranean regions, while full sun makes cacti feel at home. Shade-lovers from the rainforest floor thrive some distance from a bright window and do best in humid rooms, such as bathrooms and kitchens (see pp.120–121).

SUPPLYING EXTRA LIGHT

Artificial lights can be used where light is low. Ordinary bulbs look bright, but rarely give out enough light energy to sustain plants. Energy efficient LED "grow lights" replicate the sun's intensity and emit the colours of light that plants need to grow. Sunlight contains all the colours of the rainbow, but plants' leaves reflect green hues, and instead absorb high energy blues and violets (and invisible ultraviolet) for leaf growth, and low energy oranges and reds for triggering flowering. Commercial growers' grow lights are purple – a highly efficient combination of red and blue.

TEMPERATURE

Indoor climate control means carefully placed plants rarely face harmful temperature extremes, but summer temperatures soar on bright windowsills and in conservatories, so that all but toughest cacti and succulents may need somewhere cooler. Autumn and winter can be more problematic as heat-loving plants can start to suffer cold injury (see pp.118–119) below 18°C (65°F), which is almost guaranteed in a draughty spot or if left sandwiched between drawn curtains and the windowpane (see p.163).

A PLANT'S WATER LOSSES DOUBLE FOR EVERY

10°C
(18°F)

INCREASE IN TEMPERATURE

AT AND ABOVE

46°C
(115°F)

DEATH
IS CERTAIN

WHAT WE PERCEIVE

Green reflected
by plants
- what we see

WHAT A PLANT ABSORBS

Colours to power growth
Not all parts of the visible light spectrum are equal. Plants make use of high energy blues and violets to power leafy growth and low energy reds to trigger blooms. The green portion is reflected and visible as their leaf colour.

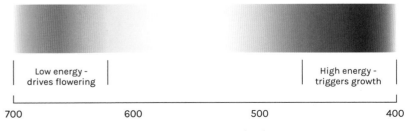

Low energy -
drives flowering

High energy -
triggers growth

700   600   500   400

WAVELENGTH OF LIGHT (NM)

GROWING ON

ARE ALL ROOTS THE SAME?

Making up around a third of a plant's total weight, roots anchor a plant to the ground, mine for nutrients, seek out water, and interact with the multitude of creatures in soil. Plants develop different roots to carry out these varied roles.

Roots are a plant's priority, and from the start, a lot of energy is invested in growing them. It is the baby root (called the radicle) that is first to emerge from the seed at the start of a plant's life (see pp.76–77). This forms the "primary root", from which one of two main types of root system will develop.

TAPROOTS AND FIBROUS ROOTS

Dig up a dandelion and you will find a long root tethering it deep into the soil. This "taproot" is the plant's primary root, from which smaller "lateral" roots arise to spread through soil. It drives straight down to form an important anchor and access water from deep in the ground. Deciduous plants also use their roots to store energy reserves through winter. Taproots are sometimes specially adapted for this purpose, as seen in the fleshy taproots of carrots.

This is very different to the root system of grasses and other members of the monocot category of plants, such as alliums, and palms, which have a dense mass of wiry roots that emerge from the seed and from the base of the stem as it grows. These are termed fibrous roots and usually spread quite close to the soil surface, where they are quick to absorb rainfall to supply the growing plant.

Roots vary in appearance but all provide the means for plants to hold firm and find water and nutrients in the soil.

FIBROUS ROOTS ARE ALL A SIMILAR THICKNESS AND USUALLY FORM CLOSER TO THE SURFACE

Extra rooting ability

PLANTS HAVE THE ABILITY TO SPROUT ROOTS FROM ALMOST ANYWHERE THANKS TO THE MERISTEM TISSUE (SEE PP.136–137) WITHIN EVERY STEM AND LEAF.
Roots that emerge from an above-ground plant part are called "adventitious" (meaning "arrived from a foreign land") and can provide stems with extra support or help plants cope in waterlogged soil. They also grow from stems that are "layered" (see p.186), or from cuttings pushed into soil (see p.185).

EXPLORING, INTERACTING, AND ABSORBING

In fact, outside desert regions, the majority of plant roots don't burrow deep into the earth. Much of the nourishment, water, and most of the living microbes are in the uppermost layers of soil, and so this is where roots concentrate their foraging.

This network is spearheaded by the tiniest "lateral roots", which measure about the width of a human hair. At their tip is an armoured "root cap" that acts a little like the head of a worm, tasting and feeling the soil with thousands of chemical receptors, twisting towards water, navigating around obstacles, and exuding a sugary slime called mucilage to lubricate its passage through soil.

Behind the root cap grow microscopic tentacles, called root hairs, which increase the surface area of root tips and absorb up to 90 per cent of a plant's nutrients and water. These structures are incredibly fragile, however, and are torn off if disturbed, potentially causing "transplant shock", where the growth of replanted or repotted plants is stunted until new root hairs become established.

Root hairs also ooze acids and chemical digesters to release nutrients from soil, as well as sugars, proteins, and, hormones to feed and attract helpful microbes (see pp.44–45). These microbes extract soil nutrients for plants in exchange for nourishment from them, while fighting off potential attackers – this bustling ring of life around roots is called the rhizosphere.

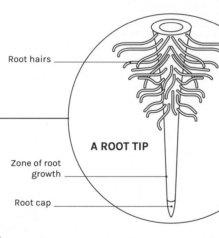

Root hairs

Zone of root growth

Root cap

A ROOT TIP

TAP ROOTS
PENETRATE DEEP INTO SOIL TO HOLD PLANTS FIRM AND SEND OUT MANY SMALL SIDE ROOTS

CAN I SIMPLY DIG UP AND MOVE PLANTS?

Moving home is particularly traumatic for plants - their roots are sliced with a spade, prised from the earth, and then sunk into foreign soil. Nevertheless, most garden plants can be moved if care is taken to minimize their ordeal.

While growing the right plant in the right place (see pp.58–59) helps plants to thrive, a garden may change over time, as large shrubs grow and cast shade or plants become overcrowded. Where this happens, relocating a plant may be in its best interests. You may also realize that a struggling plant would fare better in a different position (see pp.36–37), or you might want to redesign your garden. Herbaceous perennials tend to fare better than woody shrubs and trees when relocated – many are actually best lifted and divided regularly to keep them healthy (see pp.182–183). Trees and shrubs are best moved within five years of initial planting.

TIME A MOVE CAREFULLY

Digging up a plant is obviously traumatic, because its reduced and damaged root system not only has to heal, but also has too much aboveground growth

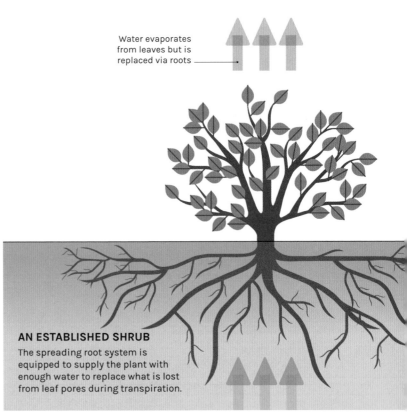

Water evaporates from leaves but is replaced via roots

The effects of transplanting

Usually, around half a plant's root system is severed when it is dug up, and the surviving roots lose many of their vital root hairs (see pp.104–105). This largely cuts off the plant's water and nutrient supply, while water continues to evaporate from leaves (see p.22) and nutrients are needed to quickly heal damaged roots. This deciduous shrub is most likely to survive if transplanted when leafless and dormant.

AN ESTABLISHED SHRUB
The spreading root system is equipped to supply the plant with enough water to replace what is lost from leaf pores during transpiration.

to supply. Growth of replacement roots is rapid, triggered by the flow of the plant hormone auxin (see pp.166–167) from stems down to roots, but this ordeal can be made easier for plants by moving at an optimal time.

Deciduous trees and shrubs are best relocated when leafless, from late autumn to late winter. This keeps water loss to a minimum and means energy reserves are safely stored in woody trunks and branches (see pp.168–169). Traditionally, branches are also pruned to help the smaller root system meet their needs in spring. But while it's wise to remove damaged branches, any more pruning adds insult to injury, triggering wounding responses and diverting sugars and nutrients away from where they're needed to repair and grow roots.

Science shows that herbaceous perennials and evergreens will recover quickly if moved when the soil is above 6°C (43°F) and moist, but leafy growth is not at its peak. This means autumn (at least five weeks before first frost) if your winters are mild and summers hot, but where winters are wet and bitterly cold, it is better to wait until spring when the soil has warmed before moving.

TIPS FOR TRANSPLANTING SUCCESS

Water plants thoroughly the day before moving and then regularly for at least three months following transplanting. Those with large taproots (see pp.104–105), including *Eryngium, Malva*, and many conifers, are difficult to transplant successfully. Most other plants send the bulk of their small "feeder" roots out sideways, so dig around and lift a wide rootball to keep as many of these as possible. Replant quickly in a hole no deeper than the existing rootball, since burying the base of a stem or trunk makes fungal infection much more likely.

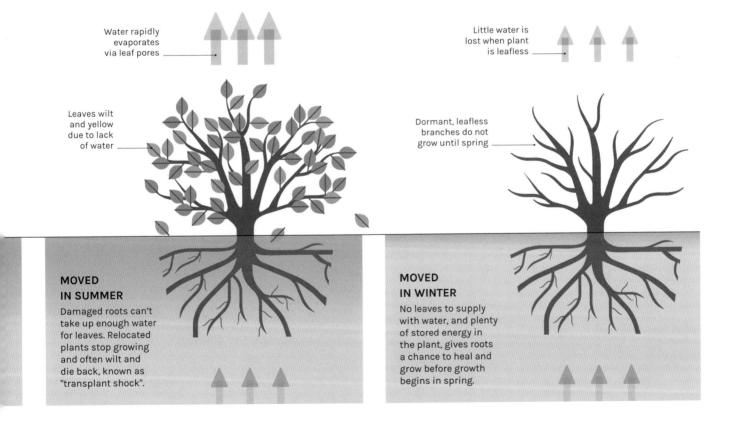

Water rapidly evaporates via leaf pores

Leaves wilt and yellow due to lack of water

MOVED IN SUMMER
Damaged roots can't take up enough water for leaves. Relocated plants stop growing and often wilt and die back, known as "transplant shock".

Little water is lost when plant is leafless

Dormant, leafless branches do not grow until spring

MOVED IN WINTER
No leaves to supply with water, and plenty of stored energy in the plant, gives roots a chance to heal and grow before growth begins in spring.

DO ALL PLANTS GET THEIR NUTRIENTS FROM SOIL?

Faced with the challenges of surviving in nutrient-poor and soil-free habitats, some plants have evolved unusual ways to glean nutrients from their environment.

NEPENTHES

THE SLIPPERY-EDGED "PITCHERS" ARE FORMED FROM MODIFIED LEAVES AND USE SCENT AND NECTAR TO ATTRACT INSECTS

PITFALL TRAP

BUGS THAT FALL IN ARE DROWNED, DISSOLVED, AND DIGESTED BY THE PLANT IN A POOL OF ACIDIC ENZYMES

The idea of plants learning to eat animals seems ludicrously unlikely, yet some plants struggling in nutrient-poor, waterlogged soils have evolved to repurpose the molecular machinery in their roots for eating meat. These plants trap insects, whose protein is so rich in nitrogen and other essential nutrients, that carnivory has evolved independently at least six times throughout the history of flowering plants. Carnivorous plants, such as Venus fly traps (*Dionaea muscipula*) and pitcher plants (*Nepenthes*), lure, capture, and digest their prey with ruthless efficiency.

PARASITES AND EPIPHYTES

Leeching off other plants is another ingenious method that has evolved at least twelve times to work around the need for soil. Parasitic plants, which include mistletoe (*Viscum album*) and dodder (*Cuscuta*), transformed their roots into tendrils called haustoria, which burrow into a host plant's tissues, siphoning off its precious water, nutrients, and sugars. The ghost plant (*Monotropa uniflora*) even taps into the networks of mycorrhizal fungi associated with trees for its food source.

Thousands of other plants, called epiphytes, find homes among the branches of others without harming them. These include orchids and "air plants" (*Tillandsia*), which need no soil, and instead use their wiry aerial roots to cling to other plants, rocks, or even cliff faces. Aerial roots are able to extract moisture directly from the air and epiphytes harvest nutrients by catching passing flecks of dust and soil in their tiny leaf hairs (called trichomes). Unsurprisingly, such plants tend to be rather slow growing.

WHAT IS HYDROPONIC GROWING?

Hydroponics (meaning "water work"), means growing plants in water instead of soil and may offer a solution to harmful intensive farming practices. It can even be set up at home.

For plants, soil ticks all the boxes, providing a firm foundation to anchor roots, a mix of minerals and organic matter to hold water and nutrients, and space for air (see pp.38–39). But soil is not essential: water lettuce (*Pistia stratiotes*), for example, floats on rivers, with its roots dangling into the water. In fact, most plants' roots can grow in water, providing it has enough nutrients (see p.122), oxygen, and a suitable pH. Hydroponics is not a new idea: the ancient Aztecs grew precious crops on rafts so that they could be floated to safety if they were attacked.

STRAIGHTFORWARD SYSTEMS

The simplest way to grow "hydroponically" is by housing plants in pots so that roots can grow into a soil-free "substrate", such as clay pebbles, perlite, vermiculite, or rockwool (fibre spun from molten rock). These pots are hooked up to a drip irrigation system delivering a solution of diluted fertilizer, and spaces between the substrate pieces allow oxygen to reach the roots. The Dutch bucket system works in this way and suits a range of plants.

There are many other ways to grow plants in water. The nutrient film technique involves securing plants above a continuous stream of oxygen- and nutrient-enriched water, which runs over their roots in a thin film. In raft hydroponics (deep water culture), plants float in, or are suspended over, a reservoir of oxygenated nutrient solution that is renewed periodically.

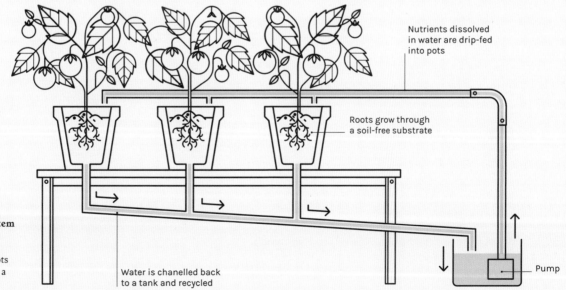

Nutrients dissolved in water are drip-fed into pots

Roots grow through a soil-free substrate

The Dutch bucket system
A simple set-up, where nutrient solution is delivered to plants in pots of soil-free substrate by a drip irrigation system.

Water is chanelled back to a tank and recycled

Pump

HOW CAN I KEEP POT PLANTS HEALTHY?

Roots love being in soil, where they can explore for water and nutrients, aided by helpful soil bacteria and fungi. Living in a container puts a stop to that and calls on the grower to meet a plant's needs instead.

Keeping your plants healthy starts with choosing the right container and compost, and continues with regular feeding, watering, and repotting.

CONTAINER CHOICE

Boots, bathtubs, buckets – plants will grow in almost any container, provided drainage holes are made in the base for water to escape. But water drains slowly from compost (see right), so even purpose-made pots may need extra holes.

Small pots dry out quickly and need watering more often, although large pots are often heavy. Research shows that the bigger the container, the larger a plant will grow. Scientists don't know how, but roots "sense" their pot's size, and limit aboveground growth to match their expected spread.

What a container is made from will influence growing conditions. Terracotta pots have a porous structure that absorbs water from compost and so need to be watered more often than plastic pots. Terracotta, ceramic, and concrete are all materials that have some air in them, insulating compost. Metal pots conduct heat rapidly, becoming dangerously hot if placed in summer sun and freezing easily in winter.

COMPOST

Peat-free multipurpose compost suits most annual plantings, but like many potting composts (see p.49) will gradually compact as the organic matter within it degrades, making it difficult for roots to absorb water and nutrients. Composts containing soil ("loam-based") have a more stable pore structure, making them the best choice for longer-lived plants. Use ericaceous compost (pH4–5) for acid-loving plants (see pp.40–41). Fill pots to about 4cm (1½in) below the rim, so that water pools on the surface and soaks into compost after watering, rather than running off the edge.

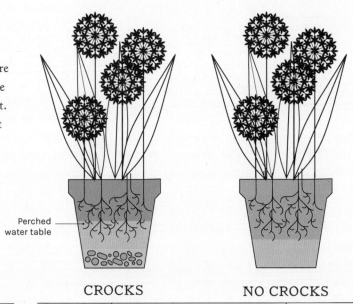

Perched water table

CROCKS

Raised water level effectively reduces the space available for healthy root growth.

NO CROCKS

Lower water level maximizes well-drained compost for healthy roots.

One crock of a myth

It's long been believed that shards of pot ("crocks"), or gravel in the base of a pot help water to drain. But thanks to surface tension (see top right), water actually pools higher up the pot.

WATERING AND FEEDING

Containers dry out quickly, so check the moisture level of compost frequently (see pp.112–113) and be prepared to water regularly – perhaps daily in summer. Give a good drenching, until water can be seen coming from the base. Avoid allowing compost to dry out too much, because it may then not absorb water easily, making effective watering difficult. From about six weeks after planting, the nutrients in compost will have been used or washed away and need to be replenished (see pp.122–123). Most can be refuelled with a liquid feed every 2–3 weeks, but the type and amount of fertilizer should be tailored to each plant. Repot plants into a larger container before growth slows or leaves yellow. For large plants where this isn't possible, remove the top 10cm (4in) of compost every year, and replace it with fresh compost.

Wet layer at base

WATER'S SURFACE TENSION MAKES IT CLING TO THE PORES within compost, causing it to pool in the bottom of the compost to form a "perched water table". If drainage is slow this wet layer can starve roots of oxygen.

Shape affects moisture levels

Since some water accumulates at the bottom of even the best drained pots, their shape influences the space for healthy root growth and the availability of moisture.

SHALLOW

Potential for waterlogging
The relatively high water table can leave roots too wet.

STANDARD

A happy medium
Roots can reach moisture and plenty of well-drained compost.

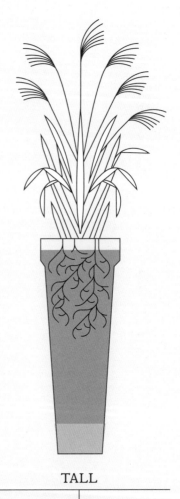

TALL

Left high and dry
Far above the water table, roots struggle to obtain water.

HOT, SUNNY, WINDY
– WITH WATERING

Water is lost through leaves and from soil. Regular watering keeps cells rigid ("turgid") and prevents wilting.

WARM, HUMID, STILL

Roots need less water to keep cells turgid when evaporation is slow in humid, still air. Less watering needed.

COOL, STILL, CLOUDY

Little moisture is lost when it's cool and still, so the demand on roots is low. Less watering needed.

Water needs change with the weather

Heat, humidity, and wind all affect how quickly water evaporates from leaves during transpiration (see p.22), how much is needed by roots, and when you need to water.

TURGID

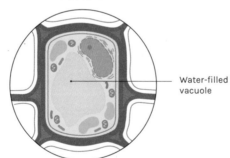

Water-filled vacuole

HOT, SUNNY, WINDY –
NO WATERING

Rapid evaporation from leaves and soil. Cells lose rigidity and shrink, and plants wilt.

Vacuole shrinks as water is lost from cell

WILTING

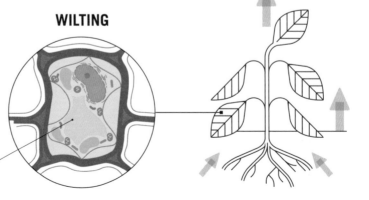

HOW OFTEN SHOULD I WATER MY PLANTS?

All too often we misjudge plants' basic need for water, dehydrating them through neglect or drowning them with love. Plants can't say when they're thirsty, but do give signs that we can read to get the balance right.

Plants lose life-giving water almost as quickly as they take it in. Water travels from soil into the roots and up the stem as a watery sap. This upward movement happens because most (95 per cent) of this water evaporates from leaves, through stomata (tiny pores mostly on their underside), drawing a continual supply up the stem like a drinking straw. Called "transpiration" (see p.22), this flow is driven by heat, humidity, and wind (see left).

PLANTS HAVE DIFFERING NEEDS

Watering routines and reminder apps are widely promoted, but are for the birds – each plant has individual needs, which change throughout its life and with the conditions. Like babies, young plants dehydrate quickly because of their small size, and are a top priority for watering, along with anything newly planted that has yet to establish its roots. For good harvests, also keep vegetables well watered.

Plants in pots need much more watering than those in the ground and should be checked daily in summer (see pp.110–111). Conversely, established perennials need little watering when planted in a suitable place (see pp.58–59).

The inbuilt adaptations of individual plant species also hold huge sway over their water needs (see pp.116–117). Those with lush, large, thin leaves lose water quickly, meaning more watering, whereas plants with small, thick, waxy, or fuzzy leaves are adapted to dry environments and are unlikely to need watering in open soil.

SOIL INFLUENCES WATERING

The type and condition of the soil also affects how often plants need watering. Clay soils store the most water (see p.39), as do potting composts with plenty of absorbent organic matter (see p.49) or vermiculite (a sponge-like mineral), while sandy soils dry out rapidly.

Unless the soil has a good structure, however, plants find it hard to take in water and can merely sip from it – microscopic root hairs can only absorb water from the tiniest soil pores (see pp.42–43). "No dig" soil that has been regularly mulched with organic matter will have pores of all sizes, which hold moisture while making it available to plants. A mulch also dramatically reduces evaporation from soil – which can be as much as 35 per cent of all water where plants are separated by bare earth – minimizing the need for watering.

RESPOND TO EACH PLANT'S NEEDS

Given all these factors, precise rules for watering are meaningless. The one golden rule is to water a plant when it's thirsty, so watch for cues from plants and the soil. Press your finger 2–5cm (1–2in) into soil – if it feels dry, then it's time to water (see pp.114–115). Lifting pots regularly also allows you to judge when they feel light and need watering.

Wilting leaves mean water pressure has already dropped in a plant, suggesting roots are dry and it's time to water. This can also be a sign of root damage or disease, however, so always check that the soil is dry.

WHAT'S THE BEST WAY TO WATER PLANTS?

Watering can confuse even experienced gardeners, perhaps partly because we don't water plants, we water the soil. Getting to know your soil and how plants drink from it makes it easier to water efficiently and keep plants happy.

Rain may fall from the skies, but plants drink from the soil. Most roots anchor a plant – it is the finest roots, covered with invisibly small root hairs, that absorb most water and nutrients (see pp.104–105). Water needs to reach these delicate tips and this takes longer than you might think: in sandy soil, it takes 20 minutes for water to seep down just 1cm (½in), while in clay soil this can take two hours.

LET PLANTS SEARCH FOR WATER

As sun beats down on soil, water quickly evaporates from the top few millimetres leaving the surface looking dry, while beneath it is moist. As more water evaporates, the shallowest roots start drying out. This is a good thing, because it triggers plants to send new roots down in search of water. Roots also track moisture as they grow, turning towards water (called hydrotropism) just as stems grow towards light (see p.12).

If soil continues to dry out, however, even deeper roots dehydrate, sending a surge of the plant stress hormone abscisic acid streaming up through the plant, causing leaf pores (stomata) to snap shut and slow further water loss. Signs of water stress,

How to water

Don't fret about the soil surface looking dry. Wait until it is dry to 2½cm (1in) before watering established plants, and half that depth for young plants. Drench soil to make moisture available to deep roots. Frequent light watering keeps roots shallow and prevents them tapping into resources deeper in the soil.

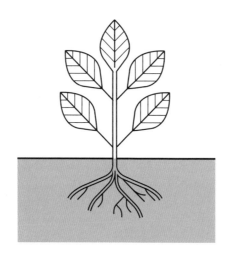

AFTER WATERING

Soil is wet
Soak soil when watering to make sure all roots can access water.

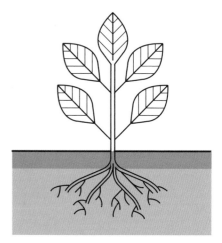

SOIL SURFACE DRY

No need to water
As shallow roots dry out, deeper roots grow and develop to reach water.

including wilting, dull-looking leaves, and stalled growth will soon follow.

SAFE AND SUSTAINABLE WATERING

Using a spray nozzle on a hose or a "rose" on a watering can create a shower of water droplets that prevents soil being washed away (eroded) and is gentle for young plants. Aim directly at the soil, because droplets showered into the air will evaporate before they get near deep roots. Avoid watering in the midday sun, because water will evaporate when it hits hot soil, rather than seeping downwards.

We live on a blue planet, but just 0.4 per cent of Earth's water is safe to drink and 98 per cent of that is found deep underground. Using a garden hose or sprinkler unnecessarily wastes this precious resource. Plants don't need clean drinking water, so wherever possible capture rainwater in butts for the garden's needs or re-use "grey water" from washing dishes or bathing, which is safe for plants as long as it doesn't contain powerful detergents or bleach.

Gardeners sometimes warn against using tap water, believing the traces of fluoride added in some areas will poison plants. Plants need some fluoride, and although too much can indeed be harmful, there is no evidence that levels in tap water trouble most plants. Some houseplants, including *Dracaena*, are sensitive to fluoride, so water them with rainwater where this is possible.

Tap water tends to be neutral, while rainwater is naturally acidic (see pp.40–41). There will also be some calcium in tap water in hard water areas. These differences are unlikely to affect plants in open soil, but acid-loving plants in pots may be better watered with rainwater. Some research suggests that letting tap water come to room temperature before watering may speed up growth, especially for houseplants from tropical regions.

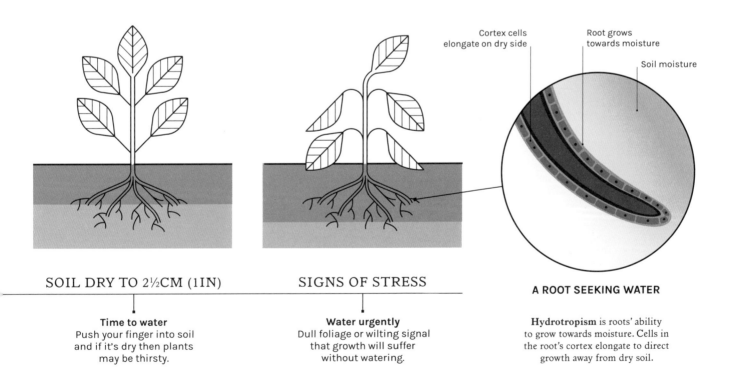

SOIL DRY TO 2½CM (1IN)

Time to water
Push your finger into soil and if it's dry then plants may be thirsty.

SIGNS OF STRESS

Water urgently
Dull foliage or wilting signal that growth will suffer without watering.

Cortex cells elongate on dry side

Root grows towards moisture

Soil moisture

A ROOT SEEKING WATER

Hydrotropism is roots' ability to grow towards moisture. Cells in the root's cortex elongate to direct growth away from dry soil.

HOW DO PLANTS COPE WITH WET CONDITIONS?

Some plants have evolved ingenious tricks to survive either partly or completely submerged in water (termed "aquatic plants") but, like the stranded sailor surrounded by water and dying of thirst, most plants struggle to cope in waterlogged soil for long.

Submerged roots absorb barely any oxygen from water, and without oxygen to breathe, lactic acid, hydrogen peroxide and other toxic chemicals build up, a little like what happens in our aching muscles when starved of oxygen during exercise. If air does not percolate into soil, the roots of most plants will die. This causes leaves and stems to wilt as they dehydrate, and the whole plant will eventually die. Some plants, however, are well equipped to handle very moist, or even sodden conditions, including dogwood (*Cornus*), hydrangeas, and meadowsweet (*Filipendula*), thanks to some clever countermeasures that can turn the tide for their roots.

ENABLING ROOTS TO BREATHE

Extra "adventitious" roots (see pp.104–105) may sprout above water level, taking over from struggling underwater counterparts. Roots in waterlogged soil can also undergo a type of internal self-destruction to form honeycomb-like spaces (called aerenchyma), which serve like tubes to transport air down from the plant above, and as exhaust pipes for carbon dioxide to escape. Roots might also create an airtight coat along their length so that oxygen in the aerenchyma can't seep out into the water. Plant species that live underwater for lengthy periods have pre-made aerenchyma and particularly airtight root linings.

Water flows through xylem tubes in leaf veins

Leaves have many stomata (pores) to release water vapour

LUSH LEAVES WITH MANY STOMATA DRAW UP LARGE VOLUMES OF WATER FROM ROOTS VIA TRANSPIRATION

Primula beesiana flourishes in damp soil, similar to that found in its native mountain meadows in China.

HOW DO PLANTS COPE WITH DROUGHT?

In our warming world, the plants that will endure are those best able to cope with hot, dry conditions. Plant species that evolved in dry regions are built to survive baking, rainless spells and their special abilities make them ideal for a dry garden.

As soon as roots detect that soil is too dry, they send a chemical distress signal (abscisic acid) upwards, telling the plant to conserve water at all costs. The breathing pores (stomata) in leaves (see p.22) snap shut to stop water evaporating away, but this cuts the supply of carbon dioxide needed for photosynthesis (see pp.70–71), leaving plants faced with the prospect of either shrivelling to a crisp or running out of food. Photosynthetic machinery tries to carry on running, but without fuel, a build-up of toxic substances slowly poisons the plant, causing leaves and flowers to fall.

WATER-SAVING ADAPTATIONS

There is therefore a big advantage if plants can find ways to mimimize water loss and cope with dry soil. Many, including pine trees and lavender, have developed small or needle-like leaves with fewer stomata, to reduce how much water can escape. Cacti have even transformed their leaves into spines, and photosynthesize and store precious water in their plump bodies.

A thick waxy outer layer (cuticle) on leaves, like those of many evergreen shrubs, is another effective barrier to water loss. Drought tolerant plants often have silver-grey foliage, which reflects intense sunlight, and many, including sage and rock rose (*Cistus*), also have a fuzz of tiny water-retentive hairs (see right). Palms send roots down 5 metres (16 feet) to plumb water rising from aquifers and so need no rain.

FINE HAIRS (TRICHOMES) TRAP MOISTURE CLOSE TO THE LEAF SURFACE TO HELP SLOW WATER LOSS

Fine hair-like appendages on the leaf surface, called trichomes

Trapped water droplets help slow moisture loss

Convolvulus cneorum is well adapted to drought thanks to its slender leaves, covered with silky, silver hairs.

WHY ARE FREEZING TEMPERATURES SO DAMAGING?

Water is the stuff of life – the molecule upon which every living thing is based (plants are 80–90 per cent water). When liquid water solidifies into ice, however, all of life's essential chemical reactions seize, turning it from life-giver into deadly killer.

Plants can be assaulted by sub-zero temperatures in two ways, termed "frost" and "freezing". Frost develops on cloudless nights when the air is still and moisture in the air freezes into tiny ice crystals on the plant's cold surfaces. Freezing happens when bitterly cold winds blow in, causing the air temperature to plummet, possibly for days rather than just overnight, and is much harder to protect against (see p.160–161).

When water within plants freezes at 0°C (32°F), it expands and forms ice crystals that are spiky and puncture cells from the inside. Plants can also suffer dehydration in prolonged sub-zero conditions, as an increasing amount of their internal water is frozen, and cells are progressively drained of their liquid.

COLD TOLERANCE ADAPTATIONS

Plants have evolved a variety of ingenious ways to survive these frozen horrors. A plant's ability to withstand cold temperatures is termed its "hardiness" (see pp.80–81). Tender plants have no defence against sub-zero temperatures and so will need protection if they are to

CAMELLIA
BLOOMS EARLY IN SPRING WHEN DAMAGE TO FROSTED PETALS IS WORSENED IF THEY THAW RAPIDLY IN MORNING SUN

Vulnerable new shoots

Young spring growth is susceptible to damage by spring frosts because the soft shoot tips haven't yet ripened and accumulated protective sugars.

have any hope of making it to spring (see pp.160–161). Half-hardy plants can withstand a light frost (down to -5°C/23°F), while those that are fully hardy are well equipped for a deeper freeze.

Hardy plants combat the ravages of ice by turning stored starches into sugar (see p.162) and producing antifreeze proteins to stop ice forming in the water between cells. They also synthesize dehydration-induced proteins to stop themselves drying out as frost sets in.

Woody species have an advantage over softer herbaceous plants because their tough bark insulates them from the cold like a timber wall on a house. Experiments have shown that when dormant, some woody plants can even withstand temperatures as low as -196°C (-321°F). Newly emerging shoots and spring blossom are soft and have no such protection, so can easily be damaged by a late spring frost, seen as brown petals, leaf edges, or shoot tips.

FREEZE-DRIED

Many evergreens are well adapted to tolerate freezing temperatures, but can nevertheless suffer the worst of winter's drying effects. Unlike deciduous plants,

which enter dormancy in winter, evergreens carry on slowly growing, pulling water and nutrients from the soil. When soil freezes, however, this water supply runs dry. Water continues to evaporate through leaves (see p.22) and the top of a plant can become dangerously dry. Combine this with the fact that cold air has little moisture, and winter winds can easily dry out a plant's extremities to a crisp. Termed "winter burn", these dying brown patches may only appear when temperatures rise. Protecting evergreen plants from the wind (see p.160–161) and watering evergreen shrubs well in autumn will help prevent winter burn. Plants with a "cushion-like" form often hail from harsh alpine habitats. Hugging the ground shelters them from freezing winds and enables their survival beyond -15°C (5°F).

When frost fades, ice may still have a final sting in its tail. If a plant that has survived freezing is warmed too quickly, the puncture holes within its cells don't have time to heal, and its internal structure can turn to slush. Given that the worst frosts happen at night, plants facing east (see pp.32–33) that are warmed by the morning sun are at particular risk of thawing damage.

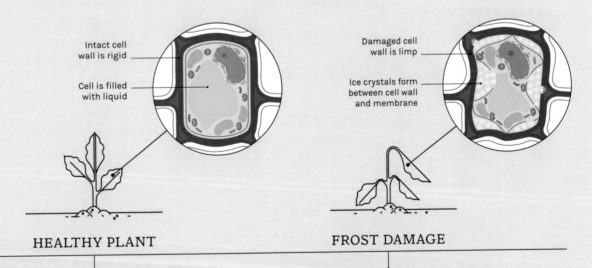

Intact cell wall is rigid

Cell is filled with liquid

Damaged cell wall is limp

Ice crystals form between cell wall and membrane

HEALTHY PLANT

FROST DAMAGE

At temperatures above freezing plant cells are filled with liquid, which keeps cells rigid (turgid) – holding the plant upright – and creating an environment where life processes can continue normally.

During freezing weather ice crystals can form within the cells of plants that aren't fully hardy. Expanding crystals puncture cell walls and membranes, so that liquid escapes when they melt, causing plant tissues to crumple.

WHAT'S THE BEST WAY TO RAISE HUMIDITY FOR INDOOR PLANTS?

Many favourite indoor plants come from subtropical regions – often from rainforest habitats where air humidity is upwards of 80 per cent. It's no surprise then that they struggle in the dry atmopshere of the average home and need a little help to flourish.

The typical humidity inside a house will be around 40 per cent, meaning that the air is holding 40 per cent of the water vapour that it could possibly contain. This is a much drier environment than the natural rainforest habitat of many houseplants and causes them to dehydrate as internal moisture quickly evaporates through "breathing pores" on the underside of leaves called stomata (see p.22). Symptoms caused by dry air include: brown leaf tips, yellowing leaf edges, leaf fall, and wilting.

WHY IS HOUSEHOLD AIR SO DRY?

Central heating causes indoor humidity to fall, because as air warms the maximum amount of moisture it can hold increases and so the air feels drier: if the humidity is 40 per cent in a room at 18°C (64°F), for example, then increasing the thermostat to 22°C (72°F) would actually drop the humidity to 31 per cent. Air also tends to be drier during winter, when indoor humidity is carried away by cold drafts from outside.

Increasing the humidity of household air to the levels of the tropics would be uncomfortable and cause condensation to trickle down walls and windows. Luckily this is unnecessary because most houseplants are happy with humidity at around 50 per cent. Lots of strategies are purported to increase the humidity around plants, but some are more effective than others so choose carefully.

THE MOST EFFECTIVE METHODS

The simplest way to keep most indoor plants happy is to house them in a bathroom, where the average humidity is around 50 per cent and may rise to over 90 per cent when a hot shower is running. In other rooms an indoor humidifier can be used to add water vapour to the air and is a failsafe strategy for keeping a room at a comfy air moisture level for

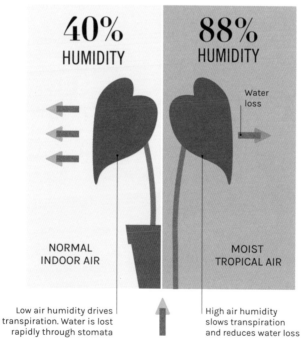

40% HUMIDITY
NORMAL INDOOR AIR

88% HUMIDITY
Water loss
MOIST TROPICAL AIR

Low air humidity drives transpiration. Water is lost rapidly through stomata

High air humidity slows transpiration and reduces water loss

WATER TAKEN UP THROUGH ROOTS

Air humidity affects water loss Inside leaves humidity is almost 100 per cent, creating a concentration gradient with drier air outside that drives transpiration.

Moth orchids: As natives of warm, humid forests in Asia and Australia, *Phalaenopsis* thrive indoors in humid air.

your green tenants. Placing an upturned jar or transparent container over a pot plant – or growing plants inside a glass vessel – creates a sealed environment, called a terrarium, where most of the water that evaporates from leaves cannot escape, increasing the humidity of the air inside. Plants in a terrarium need little watering because water vapour condenses on the glass and dribbles down into soil.

TRICKS THAT MAY HELP

All plants give out the water that they have drawn up from their roots through their leaves as vapour during transpiration (see p.22). Grouping plants together helps to capture this vapour and create a bubble of still, humid air around them.

Misting plants with a water spray is a common way to raise humidity, although its effects are very short-lived: droplets settle on leaves and gradually evaporate, briefly increasing the humidity around

MISTING

SPRAYING PLANTS WITH FINE WATER DROPLETS DOES LITTLE TO INCREASE HUMIDITY IN DRY AIR

them, before the moisture dissipates into the room. On dry winter days, water droplets may only linger for 10–15 minutes, meaning that hourly spraying will be needed to have any impact.

A POPULAR MYTH

Pebble trays are frequently recommended to keep indoor plants happy. The theory goes that evaporation from a shallow tray or dish filled with water and pebbles will increase the humidity in the air around plants standing on the pebbles. In reality, however, it does nothing to help the plant. The humid air just above the water doesn't rise straight up, but diffuses in all directions, meaning that the humidity at the height of a plant's roots and leaves is essentially unchanged.

WHICH NUTRIENTS ARE ESSENTIAL FOR HEALTHY PLANTS?

Plants take up a range of nutrients from soil through their roots, each of which is vital for strong, healthy growth. Understanding the role and importance of different plant nutrients will help you to keep plants growing happily and recognize signs of deficiency.

Until the mid-1600s, it was thought that plants "ate" soil. Only when a scientist, Jan Baptist van Helmont, found that in five years a potted willow grew by 74.5kg (164lb) when only 57g (2oz) of soil was lost, did it become clear that this is not the case. This seemingly miraculous growth occurs because plants use sunlight to make food from thin air (see pp.70–71) to power the processes within their cells.

But, like humans, plants need more than just energy and become malnourished without a range of life-essential nutrients (represented by the soil "eaten" by van Helmont's tree). These nutrients are found dissolved in soil water, which allows them to be absorbed through plant roots.

NITROGEN

Proteins for building and repairing tissues are assembled from nitrogen (N), which makes this the most important nutrient for growth and health. Nitrogen is a key part of the green, photosynthetic pigment, chlorophyll (see pp.70–71), and if it is lacking chlorophyll production stutters and leaves turn yellow ("chlorosis"). Even though nitrogen makes up 78 per cent of the air, it is difficult for plants to access. Much of soil's plant-ready nitrogen is recycled from organic matter by fungi and microbes (see pp.44–45), and there are also "nitrogen-fixing" bacteria, which extract it from the air. Legume plants, such as peas and beans, have evolved a partnership with nitrogen-fixing bacteria

("rhizobia"). These live within lumps on roots called nodules, and supply nitrogen in return for sugars.

PHOSPHORUS AND POTASSIUM

The other major (primary) nutrients, phosphorus (P) and potassium (K), come from the rocks that made the soil's mineral content (see pp.38–39), as do calcium (Ca), sulphur (S), and magnesium (Mg), which are secondary nutrients, and needed in smaller amounts. Plants struggle to obtain phosphorus, and use helpers from the soil food web to do some heavy lifting for them. The "mycorrhizal" fungi around roots (see p.45) pass traces of phosphorus from soil into roots. Plants need phosphorus to unlock the energy in food, so where it is lacking the growth of young plants is stunted and older plants stop flowering and fruiting. Potassium is vital to keep a plant's molecular cogs working – a little like salt for us. It is usually plentiful in all bar sandy and chalky soils, where it is quickly washed away.

TRACE ELEMENTS

A handful of elements are needed in only the tiniest amounts. These trace elements (tertiary nutrients) are iron (Fe), manganese (Mn), boron (B), copper (Cu), zinc (Zn), molybdenum (Mo), and chlorine (Cl). As with other soil nutrients, poor growth and yellowing or other discoloration between or around leaf veins are indicators that these vital ingredients for life are in short supply.

Plant nutrients

A variety of soil nutrients is essential for healthy and vigorous plants. Primary nutrients are required in the largest quantities, while just traces of tertiary elements are enough.

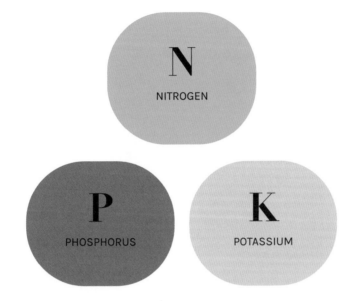

PRIMARY

Nitrogen provides building blocks for growth, phosphorous helps to access energy, while potassium keeps processes running smoothly.

SECONDARY

Root growth is dependent on calcium, magnesium is a component of chlorophyll, and sulphur is needed for energy production.

TERTIARY

Fe	Mn	Zn	Cu	Mo	B	Cl
IRON	MANGANESE	ZINC	COPPER	MOLYBDENUM	BORON	CHLORINE

Trace elements are involved in growth, energy production, enzyme and hormone creation, and the conversion of nitrogen to accessible forms.

WHAT'S THE BEST WAY TO FEED PLANTS?

Plants are adept at making their own food from sunlight, air, and water (see pp.70–71), but to put on healthy growth they also need a range of life-essential nutrients to be available in soil (see pp.122–123). It is these that gardeners need to replenish.

Manufacturing fertilizers for garden plants is a multibillion-dollar industry, yet in nature plants thrive because nutrients are recycled when leaves fall, plants and animals die, and creatures poop. The soil food web (see pp.44–45) is the ultimate recycling machine: one creature's faeces is another's breakfast; one dead carcass is a feast for a thousand tiny mouths. Nothing is wasted and all the soil nutrients upon which plants depend, including nitrogen, phosphorus, potassium, and calcium, are replenished in forms ready for roots to take up.

FEED YOUR SOIL NOT YOUR PLANTS

Far better than feeding your plants is to feed your soil and support this natural nutrient recycling system. This is easy to achieve by adding an annual mulch (layer) of well-rotted compost to the soil surface, usually in late autumn or spring. This could be homemade compost (see pp.188–191), or a bought version made from green waste, spent mushroom compost, or other organic materials.

You don't need to dig compost into soil, because laid on the surface it feeds the soil food web just as well and protects soil from erosion during winter weather. Research also suggests that digging disrupts the complex networks of soil life, which slowly turn your mulch into perfect plant food, and convert nutrients into forms not easily washed away by heavy rain (see pp.42–43). Decaying leaf litter replenishes nitrogen and phosphorus, while nitrogen is boosted when plants in the legume family, such as clover, die and decompose into soil, as "green manure".

Advantages of feeding the soil

Organic matter is processed by soil life to create a nutrient store that plants can access when needed, with the help of mycorrhizae. This fungal partnership is damaged when synthetic fertilizers supply short-term bursts of nutrients, making plants less able to extract what they need from soil and more reliant on fertilizers.

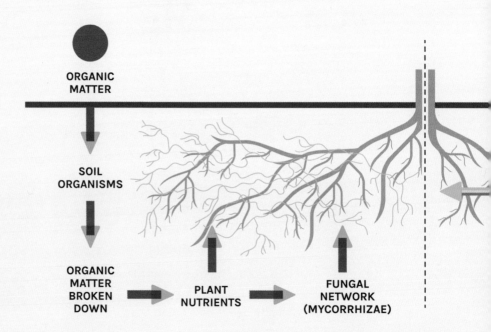

ORGANIC MATTER

SOIL ORGANISMS

ORGANIC MATTER BROKEN DOWN

PLANT NUTRIENTS

FUNGAL NETWORK (MYCORRHIZAE)

CAREFUL FERTILIZER USE

The concentrated nutrients contained in fertilizers undoubtedly boost growth, but give a short, fast-acting hit, like a handful of sugary candy. Feeding plants directly involves anticipating their needs and judging quantities of nutrients, which is difficult to do, even with years of experience. Adding an excess of one nutrient can lead to deficiencies in others, while any fertilizer use risks upsetting the delicate soil food web, including the "mycorrhizal" fungi that help roots gather water and nutrients (see pp.44–45). The nutrients in synthetic fertilizers also dissolve very easily and so quickly wash away in heavy rain, polluting rivers and streams (see pp.30–31).

Plants in containers are limited to the resources available in the compost within their pot and so are reliant on you to replace the nutrients they use. Fresh potting mix generally supplies nutrients for 4–6 weeks. After that regular feeding is essential, although how often will vary according to the pot size and the needs of individual plants. Most houseplants need feeding once or twice a month when in growth, while tomato plants thrive when fed twice a week. Add granules of slow release fertilizer to compost to deliver nutrients over an entire growing season.

Fertilizers

THE AMOUNTS OF THE THREE PRIMARY NUTRIENTS ARE USUALLY GIVEN AS AN "NPK RATIO" ON THE PACKET, WHICH INDICATES THE PERCENTAGE OF NITROGEN (N), PHOSPHORUS (P) AND POTASSIUM (K).
A 100g (4oz) bag of 7:7:7 fertilizer, for example, would contain 7g (¹/₄oz) of each major nutrient – although as the solid minerals aren't pure, plants don't actually get the full 7g (¹/₄oz) of potassium and phosphorus. Fertilizers can be either synthetic or organic, the latter coming from animal or plant sources rather than a chemical factory. The nutrients in synthetic fertilizers are in highly water soluble forms, instantly available to plants, but will likely seep into and through the soil much faster (unless manufactured in a slow-release form). Organic fertilizers are often slower to act, but remain in the soil and available to plants for longer.

SYNTHETIC FERTILIZER
Nitrogen, phosphorus, potassium

PLANT NUTRIENTS

NUTRIENT LEACHING

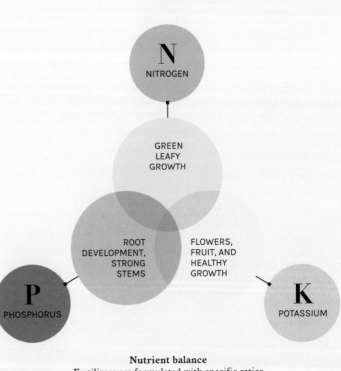

Nutrient balance
Fertilizers are formulated with specific ratios of the primary nutrients (NPK) to suit different purposes. Those high in nitrogen, like spring lawn feed, promote leafy growth, while those with more potassium, including tomato feed, encourage flowers and fruit.

WHAT'S THE SECRET TO A HEALTHY, ATTRACTIVE LAWN?

A lush lawn needs regular care and attention, but this needn't mean setting aside environmental concerns. Getting to know your grass is the first step to helping it thrive, and just might give you more time to relax and enjoy it.

With 100,000 sun-loving blades per square metre (10.8 square feet), the first secret to a happy lawn is light. Grasses are the most efficient plants at turning sunlight into food, with longer leaf pores (stomata) than broad-leaved plants, which let carbon dioxide and oxygen flow in and out faster (see pp.70–71). Where light is low, prune back encroaching trees and select shade tolerant turf mixes.

A STEADY FOOD SUPPLY

The rapid growth of grass gives it a huge appetite for nitrogen – a key building block of chlorophyll and proteins. To satisfy this, gardeners often apply fast-acting high-nitrogen fertilizer ("lawn feed"). However, about half of this nitrogen simply washes through soil to pollute streams and groundwater.

A more environmentally conscious alternative is to spread a thin 1cm (¼in) "top dressing" of fine-textured compost every spring and autumn. This has the same benefits as mulching garden soil (see pp.42–43), feeding the soil food web, making nutrients available, and improving soil structure, which reduces the need for watering. Leaving short grass clippings on a lawn after mowing also returns nitrogen to the soil, satisfying about 30 per cent of its needs. But top dressing won't instantly green-up a lawn like fertilizer, because soil life has to work on organic matter to make nitrogen available.

Spiking or "aerating" creates little holes that help water, air, and organic matter work into the soil. This can be done once a year (before top dressing) using a purpose-built aerator or a garden fork, plunged 10–13cm (4–5in) deep, every 15cm (6in).

A build-up of dead grass (called "thatch") at the base of plants is often a sign that a lawn is being overfed or overwatered. Thatch is best dealt with by helping soil microbes to break it down, through aeration and avoiding synthetic fertilizers and weedkillers, which harm helpful bacteria and fungi. This will avoid the need to rake thatch off (termed "scarifying").

MAKE WATERING WORTHWHILE

Grasses are thirsty – their vast leaf area can drain lawn soil of water four times faster than bare ground. They are also resilient, however, and although a lawn may turn straw-coloured in dry weather, well tended grass will quickly green-up when rain falls. This means that most watering in temperate climates is simply to keep lawns looking good, rather than for their health.

If you do water, then simulate natural rainfall with a prolonged, thorough soak once or twice a week. Brief waterings don't permeate far into soil, keeping roots shallow, and ultimately make plants more likely to dry out.

Grass length

LONGER GRASS STAYS GREENER
Grasses grow from a "crown" near ground level. Mowing too low can injure this growing point and delay a lawn's recovery. Cut slightly higher, at 4-5cm (1½-2in), grasses root deeper and are able to stay greener in summer.

A healthier, more resilient lawn

Aerating and top-dressing a lawn with compost may seem like hard work, but it will develop thick, healthy grass that looks good year-round without applying chemical fertilizer.

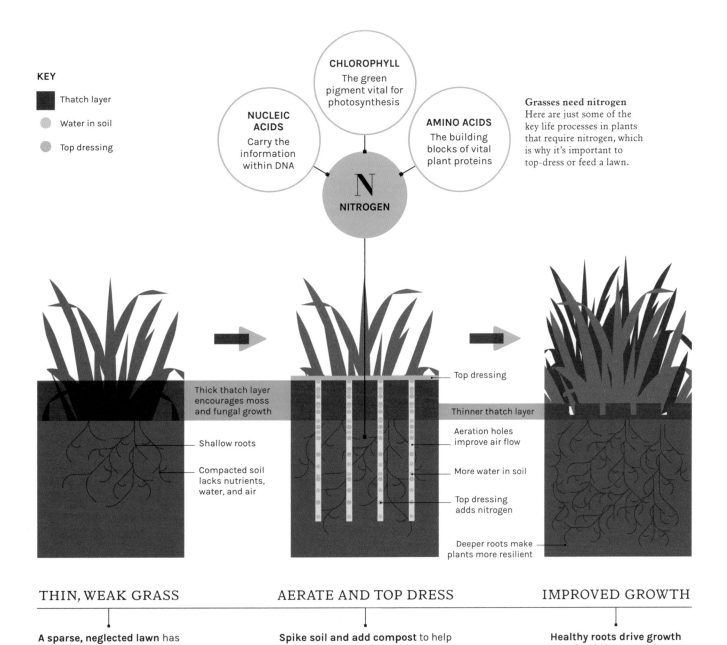

KEY

- Thatch layer
- Water in soil
- Top dressing

CHLOROPHYLL
The green pigment vital for photosynthesis

NUCLEIC ACIDS
Carry the information within DNA

AMINO ACIDS
The building blocks of vital plant proteins

N
NITROGEN

Grasses need nitrogen
Here are just some of the key life processes in plants that require nitrogen, which is why it's important to top-dress or feed a lawn.

Thick thatch layer encourages moss and fungal growth

Shallow roots

Compacted soil lacks nutrients, water, and air

Top dressing

Thinner thatch layer

Aeration holes improve air flow

More water in soil

Top dressing adds nitrogen

Deeper roots make plants more resilient

THIN, WEAK GRASS

A sparse, neglected lawn has thick thatch and shallow roots so leaves yellow in dry spells and are prone to disease.

AERATE AND TOP DRESS

Spike soil and add compost to help air and water reach roots, and to add nutrients for release by soil microbes, which also break down thatch.

IMPROVED GROWTH

Healthy roots drive growth so thick foliage crowds out weeds, and roots reach deeper to access moisture in dry weather.

HOW DO PLANTS CLIMB?

Climbing plants don't play fair, hoisting themselves upwards to reach sunlight, then flaunting their flowers and fruits to attract pollinators and dispersers, and smothering competitors. The ingenious ways they do this are truly incredible.

Runner beans, honeysuckle (*Lonicera*), and wisteria are all twining climbers that simply spiral around a support. A twiner finds a support by swirling its growing stem in the air (called circumnutation), like the blade of a windmill. When the swinging stem strikes an object, the outer surface starts growing faster than the inside (termed thigmotropism), corkscrewing the entire plant around the object so that it can pull itself up (see right).

COILING, STICKING, AND SPRAWLING

Other plants, such as grape vines (*Vitis*), climb using tendrils, which act like grasping hands. Tendrils are leaves or stems that have evolved an elongated, wiry shape, and may be branched, like those of peas. They also circumnutate until they touch a support, when they form a hooked tip and start to coil around whatever they have grasped, like

a python constricting its prey. Turning anticlockwise at one end and clockwise at the other, tendrils twist into a double-ended spring-like structure that tightens when pulled. This spring then contracts, drawing the plant closer to the support.

Some plants with tendrils, such as Boston ivy (*Parthenocissus tricuspidata*), also have adhesive pads at the tips. These produce a strong glue that sticks the tendril to a surface. Other plants, such as clematis, grasp using modified leaf stalks. Like tendrils, they will twist around anything they touch, but are too short to coil around thick objects.

English ivy (*Hedera helix*) and climbing hydrangea are tough climbers, which sprout dense clusters of short "adventitious" roots (see pp.104–105) from their stems. These work their way into tiny fissures and crevices and then expand to fill the gap. In ivy they also also sport microscopic hooks on the tips

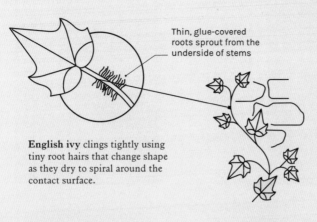

Thin, glue-covered roots sprout from the underside of stems

English ivy clings tightly using tiny root hairs that change shape as they dry to spiral around the contact surface.

CLIMBING METHODS

Garden plants use a variety of strategies to grasp supports and pull themselves upwards. The way each does this determines the type of support it needs (see pp.130–131).

STICKING

ENGLISH IVY (*Hedera helix*)

CLIMBING HYDRANGEA (*Hydrangea anomala*)

TRUMPET VINE (*Campsis radicans*)

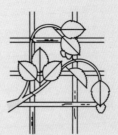

TENDRILS

SWEET PEA (*Lathyrus*)

PASSION FLOWER (*Passiflora*)

GRAPE (*Vitis*)

SPRAWLING

CLIMBING AND RAMBLING ROSES (*Rosa*)

BLACKBERRY (*Rubus*)

TWINING

CLIMBING BEAN (*Phaseolus*)

JASMINE (*Jasminum*)

HOP (*Humulus*)

MORNING GLORY (*Ipomoea*)

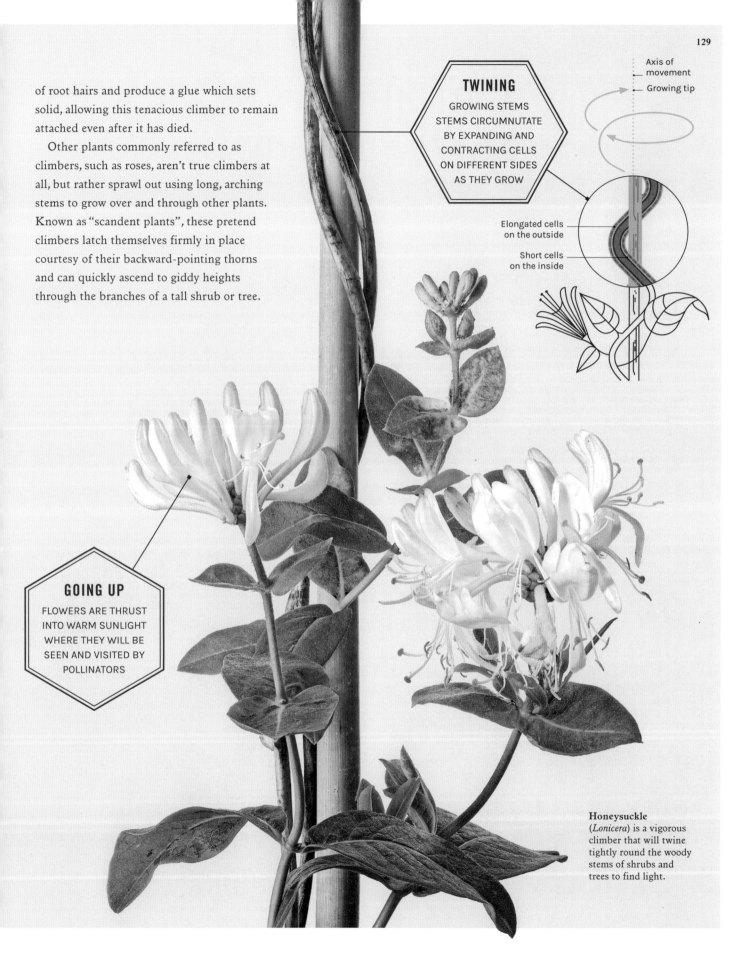

of root hairs and produce a glue which sets solid, allowing this tenacious climber to remain attached even after it has died.

Other plants commonly referred to as climbers, such as roses, aren't true climbers at all, but rather sprawl out using long, arching stems to grow over and through other plants. Known as "scandent plants", these pretend climbers latch themselves firmly in place courtesy of their backward-pointing thorns and can quickly ascend to giddy heights through the branches of a tall shrub or tree.

TWINING
GROWING STEMS STEMS CIRCUMNUTATE BY EXPANDING AND CONTRACTING CELLS ON DIFFERENT SIDES AS THEY GROW

Axis of movement
Growing tip

Elongated cells on the outside

Short cells on the inside

GOING UP
FLOWERS ARE THRUST INTO WARM SUNLIGHT WHERE THEY WILL BE SEEN AND VISITED BY POLLINATORS

Honeysuckle (*Lonicera*) is a vigorous climber that will twine tightly round the woody stems of shrubs and trees to find light.

WHICH PLANTS NEED SUPPORT?

Although some plants can scale a surface better than a gecko, a surprising number benefit from a helping hand to stay upright. Knowing what and when to support will help plants to perform at their best and prevent them being damaged.

Soft-stemmed annual and perennial plants with tall, flamboyant flowers, including sunflowers and dahlias, may need support to prevent them toppling in high winds or summer downpours, especially if they have been raised under cover and their stems have not toughened during exposure to outdoor conditions (see pp.86–87). Staking stems by loosely tying them to an upright cane or pole, pushed firmly into the soil works well.

KEEPING PERENNIALS UPRIGHT

Other herbaceous perennials with large blooms, such as double-flowered peonies, get weighed down when wet, snapping or bending their stems, and do better with supporting stakes or girdles. Shore up tall herbaceous perennials, such as delphiniums, with a horizontal grid, put in place on sturdy wire legs during spring, for their stems to grow through. Hidden by foliage, these keep plants upright during

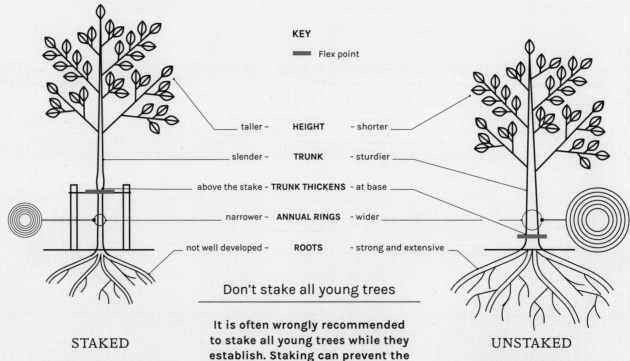

KEY

▬ Flex point

taller –	**HEIGHT**	– shorter
slender –	**TRUNK**	– sturdier
above the stake –	**TRUNK THICKENS**	– at base
narrower –	**ANNUAL RINGS**	– wider
not well developed –	**ROOTS**	– strong and extensive

Don't stake all young trees

It is often wrongly recommended to stake all young trees while they establish. Staking can prevent the growing trunk strengthening as it flexes, and may hamper formation of sturdy roots to anchor the tree.

STAKED

Trees are more at risk of snapping or falling when the stake is removed. Only stake in exposed locations.

UNSTAKED

Unsupported trees become shorter and stockier, with a trunk that is wide at the base to resist wind damage.

Selecting plant supports

To find the right support for each plant, carefully consider its growth habit and location, and put any structure in place before planting or spring growth.

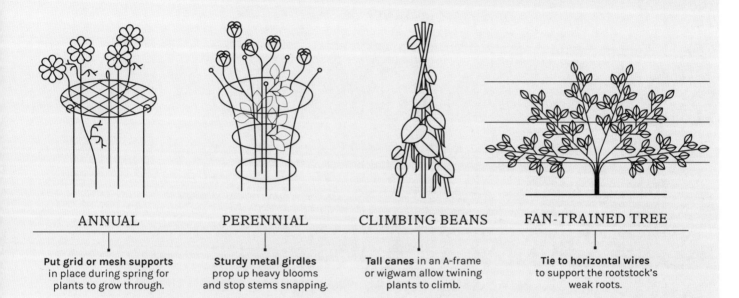

ANNUAL	PERENNIAL	CLIMBING BEANS	FAN-TRAINED TREE
Put grid or mesh supports in place during spring for plants to grow through.	**Sturdy metal girdles** prop up heavy blooms and stop stems snapping.	**Tall canes** in an A-frame or wigwam allow twining plants to climb.	**Tie to horizontal wires** to support the rootstock's weak roots.

strong winds. Perennials that form clumps of stems, like sedums and asters, can be propped up at their edges with a girdle to stop stems falling outwards.

HELP CLIMBERS PERFORM

For climbers, your choice of support will depend on the climbing style of the plant (see pp.128–129). Those that twine their main stems around supports, such as honeysuckle and runner beans, are best given tall canes or vertical wires to spiral up. Plants with tendrils, such as vines and sweet peas, grip netting, fine trellis, or twiggy sticks, as do clematis with their twisting leaf stalks. Climbing roses can be trained onto sturdy horizontal wires or trellis by tying their young stems in place. All of these will also scramble with ease through other plants. Self-clinging climbers, including ivy and *Campsis*, naturally grip the textured surface of a wall, fence, or tree trunk.

PROPS FOR YOUR CROPS

Many vegetable plants struggle without support, particularly when they bear heavy crops. Today's varieties have been bred to produce large quantities of over-sized fruits, often causing plants to buckle under their weight. Tall tomato and cucumber plants need stakes, sturdy canes, or wires to support them, while even heavier melons may need each of their fruits secured in a net.

Fruit trees intended for small gardens or for training as cordons, fans, or espaliers, are grafted onto dwarfing or semi-dwarfing rootstocks (see pp.90–91) to limit their growth. These weak-growing rootstocks form small root systems that are not enough to anchor the tree alone so, unlike other trees, their trunks must be permanently staked or the tree planted against a structure where the branches can be trained along supporting wires.

FLOWERS
AND FRUITS

HOW DO PLANTS REPRODUCE?

All life is designed to repoduce itself. Male and female animals unite their sperm and eggs, producing young that combine their parents' DNA. Plants follow less rigid rules, and have a far weirder and more wonderful sex life than you might imagine.

Most plants are "hermaphrodites", meaning they have both male and female reproductive anatomy. They also often have many sets of sex organs within numerous flowers. Each flower usually contains both male and female parts (bisexual or "perfect" flowers), but some plants, such as holly (*Ilex*) and *Ginkgo biloba* have only male or female flowers on individual plants (termed unisexual or "imperfect" flowers) and so need to find a mate of the opposite gender to produce fertile seeds.

While animals keep their eggs and sperm safely inside their bodies, plants like to flaunt their wares. Tiny round or rugby ball-shaped pollen grains containing the male sex cells are held in bright plump cushions (anthers) atop of thin stalks (called filaments). Plant eggs are nestled within the base of a flower's female central pillar (carpel).

THE POLLINATION PROCESS

Given that they cannot move, plant copulation rarely starts with a loving embrace. Rather, plant sperm (pollen) must make a leap into the unknown by either floating in the air in the hope of hitting the sticky tip of a female carpel (wind pollination) or by hitching a ride on an animal pollinator. Grasses and some trees opt for the scattergun approach of wind pollination, while other plants have evolved colourful flowers, laden with sweet nectar, to attract insects, birds, and even mammals in the hope that their sperm will be couriered courtesy of an unwitting animal carrying sticky pollen grains to nearby flowers as it feeds. This transfer of pollen between plants is called cross-pollination and combines the DNA of the parents, producing seeds that are a genetic blend of both plants.

When pollen and egg have fused (see diagram below), the flower's job is done, and its petals shrivel and fall. As the pregnant flowerhead nurses its embryos into seeds it often inflates into a protective fruit, drawing up nutrients and sugar reserves from the plant to fuel its expansion. If you care not for the fruit nor the seeds of your plants then snipping off aborts their development, saving energy, but forcing the plant to produce new blooms (see p.148).

From flower to seed

Plants invest enormous energy to ensure that flowers are successfully pollinated and viable seeds produced. This sequence reveals the basics of plant reproduction.

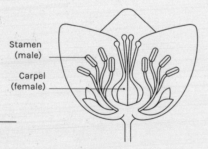

Stamen (male)
Carpel (female)

THE FLOWER

Flowers lure in insect **pollinators** using colour, scent, and the promise of energy-rich nectar.

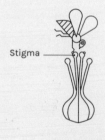

Anther
Nectary

POLLINATOR

A visiting bee is forced to brush past the anthers to reach the nectaries and its body is covered in pollen.

Stigma

POLLEN EXCHANGE

The bee carries pollen to the next flower where it brushes off onto the sticky stigma.

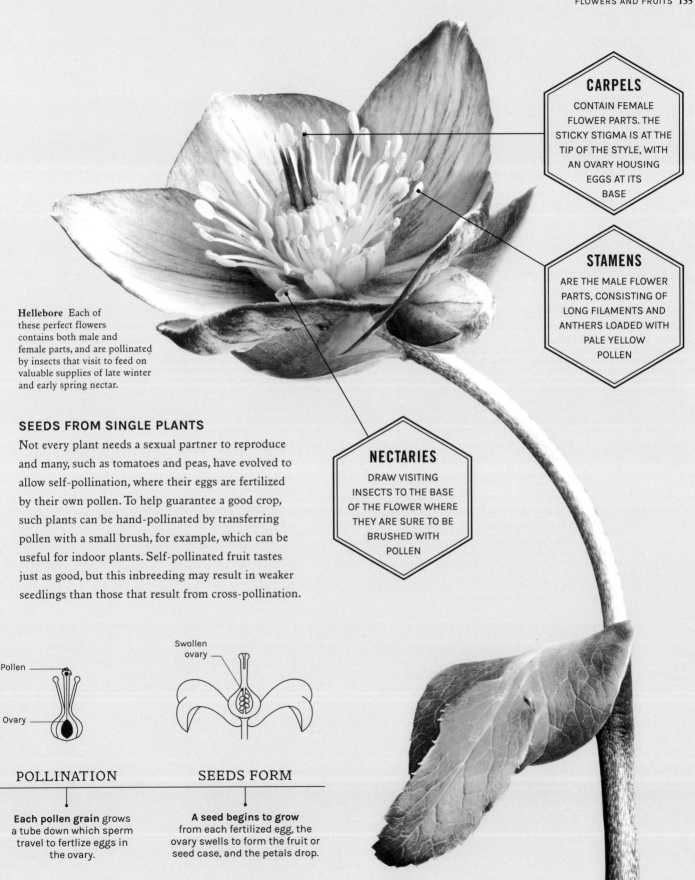

CARPELS
CONTAIN FEMALE FLOWER PARTS. THE STICKY STIGMA IS AT THE TIP OF THE STYLE, WITH AN OVARY HOUSING EGGS AT ITS BASE

STAMENS
ARE THE MALE FLOWER PARTS, CONSISTING OF LONG FILAMENTS AND ANTHERS LOADED WITH PALE YELLOW POLLEN

Hellebore Each of these perfect flowers contains both male and female parts, and are pollinated by insects that visit to feed on valuable supplies of late winter and early spring nectar.

NECTARIES
DRAW VISITING INSECTS TO THE BASE OF THE FLOWER WHERE THEY ARE SURE TO BE BRUSHED WITH POLLEN

SEEDS FROM SINGLE PLANTS

Not every plant needs a sexual partner to reproduce and many, such as tomatoes and peas, have evolved to allow self-pollination, where their eggs are fertilized by their own pollen. To help guarantee a good crop, such plants can be hand-pollinated by transferring pollen with a small brush, for example, which can be useful for indoor plants. Self-pollinated fruit tastes just as good, but this inbreeding may result in weaker seedlings than those that result from cross-pollination.

Pollen

Ovary

Swollen ovary

POLLINATION

Each pollen grain grows a tube down which sperm travel to fertlize eggs in the ovary.

SEEDS FORM

A seed begins to grow from each fertilized egg, the ovary swells to form the fruit or seed case, and the petals drop.

HOW DO PLANTS REPRODUCE WITHOUT SEEDS?

Only after decades of research were scientists able to clone an animal – Dolly the Sheep – a genetic replica of her mother, and the result of reproduction without sex. Yet for plants, this is an ordinary everyday occurence.

Plants use seeds and spores to scatter their offspring far and wide into new lands, but this strategy is always a gamble. Many plants are also able to make entire copies of themselves as a sort of insurance in case their seeds are eaten or find no fertile ground. When willow trees, for example, lose a snapped branch that falls into a river, it can float downstream and take root on a distant bank to become a whole new tree! This is termed "vegetative propagation" and by taking time to understand a little of the science behind this extraordinary ability, gardeners can harness plants' regenerative powers to turn one plant into many more.

MERISTEM OF LIFE

Every living creature bigger than a bacteria – including you – started out as a single cell, the tiniest unit of life. That one cell multiplied countless times to produce different body parts. The original cell that generates an entire living organism is called a "stem cell", since it is the stem from which all others derive. Fully grown humans have very few stem cells left in our bodies, hence we cannot, for example, regrow limbs. Plants, on the other hand, have plenty of stem cells all over their bodies, grouped together in patches called "meristem tissue". Given the right conditions, meristem cells

GENETIC CLONE

STRAWBERRY PLANTS
RAISED FROM STOLONS
SHARE AN IDENTICAL
SET OF GENES WITH
THE PARENT PLANT

will turn into roots or shoots and even develop
into entire new plants, and this is what allows a
new willow tree to grow from a broken off branch.

VEGETATIVE PROPAGATION

A huge range of plants can reproduce asexually
just like the willow, and as gardeners we harness this
capability in different methods of plant propagation,
most commonly through cuttings. By snipping off
a stem, leaf, or bud and planting it into damp soil,
a plant hormone (auxin) will seep down from the
growing tip (see p.167) to trigger the meristem
nearest the soil to generate roots. This then starts
the process for the rest of the stem to eventually
develop into an entire plant (see pp.184–185).

Some plants, such as strawberry and spider plants,
evolved to grow special long, smooth stems, called
runners (or "stolons"), which will sprout fresh
roots and birth a new plant from a bud whenever
they touch the ground. For plants such as these,
a gardener can force a new plant to sprout by
deliberately laying a runner on the surface of the
soil and securing it in place. We can even trick many

Strawberry runners
are stolons with offsets
(or plantlets), along
their length. Once
these touch soil they
develop roots and
can then survive
without a parent plant.

plants without stolons into
sprouting offsets, using a propagation
technique known as "layering" (see p.186).

Bulbs and corms (see pp.96–97) clone themselves
as they grow by budding mini-bulbs and corms off
their side, which can be broken off and replanted
(see p.187). Tubers and rhizomes also extend their
reach by enlarging their underground storage organ
sideways and sending up fresh shoots from new buds
(sometimes called "eyes"). Clump-forming plants
similarly grow outwards, making copies of each
stem (see pp.182–183).

PLANT MERISTEMS
are tiny, rapidly dividing
packets of stem cells
located at the tips of
shoots and roots, as well
as around stems within
the green layer just
beneath the bark, and
at the junctions between
leaves and stems (nodes).
The unlimited division of
these cells allows plants
to grow, heal, and
reproduce throughout
their lives.

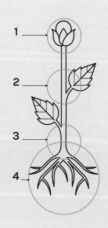

1.
**SHOOT APICAL
MERISTEM**

Extends the
main growing shoot
lengthening stems
and branches.

2.
**INTERCALARY
MERISTEM**

Allows growth
of shoots or roots
at the junction of
stem and leaf.

3.
**LATERAL
MERISTEM**

Increase the girth
of a stem and are
the source of root
growth in cuttings.

4.
**ROOT APICAL
MERISTEM**

Permits roots to
grow from their tips
and extend through
the soil.

WHY ARE FLOWERS DIFFERENT COLOURS?

From the gloriously vivacious yellow sunflower, to the alluring blue and golden hues of an iris, and the crisp white contours of an arum lily, flowers exist to attract attention and their colour is one part of how they communicate to the world.

———————

Once upon a time, all flowers were pale and colourless. Over thousands of years, plants evolved pigments to brighten up their blooms and attract pollinators – bees, wasps, birds, and bats – so that they could reproduce more effectively in increasingly cluttered landscapes. Flowers honed their hues to appeal to particular pollinators: yellows and purples for bees, pinks for butterflies, red and orange for birds, yellows for wasps and flies, with white and faded yellows reserved for the poor-sighted flies and beetles.

POLLINATOR LANDING STRIPS

Bees possess eyes that are highly attuned to purple, violet, and – unseen to us – ultraviolet light. Like most other insects, bees can't see yellow and red very well. Rather, hidden among bees' favourite yellow flowers, such as coneflower (*Rudbeckia*), common dandelion (*Taraxacum officinale*), and sunflower (*Helianthus*), are ultraviolet patterns which we are oblivious to, but which serve as a "bull's eye" to guide them into the pollen prize. Unimaginably small bumps on the surface of petals are arranged in patterns to either reflect or absorb powerful UV rays, creating these hidden patterns as a secret sign for bees and other pollinators, such as butterflies.

A CHANGING MESSAGE

Many plants are even able to communicate with pollinators in real time through changes in the colour of their flowers. Rangoon creeper (*Combretum indicum*), for example, first unfurls its chalky white petals in the evening, offering an easy target for overnight moths, before giving them a pinkish and red glow in the morning to draw in daytime butterflies and bees. Flowers of the shrub yellow sage (*Lantana camara*) darken after they have been pollinated, signalling to passing insects to feed only on the paler, unfertilized blooms. Blue and silver lupins similarly use their petals like traffic lights, turning some of their flowers' upper petals from white to magenta after pollination to prevent unnecessary visits.

BUTTERFLIES AND BEES

BIRDS

BEES

COLOUR CHOICES
DIFFERENT COLOURS ATTRACT DIFFERENT POLLINATORS

BIRDS

BEES, WASPS, AND FLIES

BEETLES, BATS, FLIES, AND MOTHS

FLIES AND BEETLES

Colour palette Insect and bird day foragers are drawn to bright colours, whereas white and pale yellow stand out in the gloom to attract night foragers and the visually challenged.

HUMAN VIEW

PETALS MAY APPEAR ONE
COLOUR TO HUMAN EYES
BECAUSE WE CANNOT
DETECT ULTRAVIOLET
(UV) LIGHT

BEE'S VIEW

PATTERNS CREATED
WHERE PETALS ABSORB
OR REFLECT UV, DIRECT
INSECTS TO THE SOURCE
OF NECTAR

PALER TARGETS

Bright colours aren't everything, though.
Large, pale blooms may be intended for the
eyes of beetles, bats, or moths. Nocturnal
moths love pale-coloured flowers
that are more visible at night and
emit strong fragrance, such as
honeysuckle (*Lonicera*). Beetles
have poor eyesight, are awkward
fliers, and need an obvious, sturdy
landing pad. Hence large, pale flowers,
such as the blooms of magnolias and water lilies
(*Nymphaea*), are particularly tempting for these insects.

An evening primrose
is not yellow to moth and bee
pollinators. Instead its pale petals
darken at the flower's centre, pointing
insects to nectaries, past stamen
loaded with pollen.

WHICH PLANTS ARE BEST FOR POLLINATORS?

A third of insect species are endangered, ravaged by the combined effects of habitat loss, climate change, pollution, pesticide use, diseases, and invasive species. But fill your garden with the right flowers and it can provide a lifeline for our insect friends.

For over 80 per cent of flowering plants, pollination – the act of moving plant sperm (pollen) onto a flower's female reproductive parts – is performed by pollinators (see pp.134–135). These are usually small flying insects, such as bees, flies, wasps, beetles, moths, and butterflies. Chief among these are bees, which pollinate a wider variety of plants than others. Far from being a selfless act, pollinators are rewarded for their services with food, drinking in droplets of sweet nectar produced by tiny nectaries at the flower base or by eating protein-packed pollen or harvesting it for their young.

SINGLE

SIMPLE FLOWERS, LIKE A SINGLE ROSE, ARE RICH IN POLLEN AND NECTAR THAT IS EASY FOR INSECTS TO ACCESS

DOUBLE

EXTRA PETALS IMPEDE ACCESS TO POLLEN AND NECTARIES AND HAVE OFTEN MUTATED FROM POLLEN-LADEN STAMENS

Double flowers The nectaries and pollen-producing parts of flowers can mutate into extra petals. Plant breeders deliberately select for this to produce double forms, but these are best avoided to maximize food for pollinators.

Flower forms

All nectar- and pollen-rich flowers are valuable for hungry insects, but flowerheads are sought-after because they are built to supply more food in each visit than single flowers.

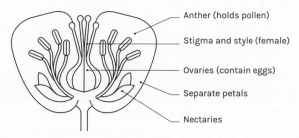

Anther (holds pollen)

Stigma and style (female)

Ovaries (contain eggs)

Separate petals

Nectaries

A single flower (hellebore) has many stamens and nectaries, but insects must fly between flowers more often.

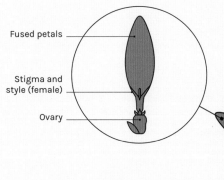

Fused petals

Stigma and style (female)

Ovary

RAY FLOWER

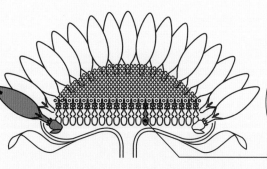

A flowerhead (sunflower), also called a composite head, contains hundreds of individual disc and ray flowers, each of which supplies insects with food.

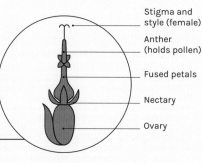

Stigma and style (female)

Anther (holds pollen)

Fused petals

Nectary

Ovary

DISC FLOWER

FLOWER QUANTITY IS WHAT MATTERS

But when it comes to satisfying the grumbling tummies of pollinators, not all flowering plants are created equal. Generally, flower number is more important than size, and plants with abundant small flowers can be visited efficiently for high returns of nectar and pollen, and typically feed the greatest number of pollinators. These include many annuals, biennials, and herbaceous perennials, but flowering trees, shrubs, and climbers, not forgetting those planted as hedges, are also some of the most significant sources of nectar and pollen in gardens.

Plants often seen by gardeners as "weeds" (see pp.46–47) also often provide a goodly fill for pollinators, such as dandelion (*Taraxacum officinale*), thistle (*Cirsium arvense*), borage (*Borago officinalis*), and common knapweed (*Centaurea nigra*), which is invasive in North America.

Some flowers, such as black-eyed Susan (*Rudbeckia fulgida*), are actually "composite flowerheads" that effectively offer hundreds of flowers for the price of one (see above). These nectar- and pollen-packed heads are rated highly by insect pollinators because they are such an efficient way to feed, saving the need to flit between many blooms. Plants such as these belong to the sunflower family (Asteraceae), which includes daisy- and thistle-shaped blooms. Each petal actually represents an individual one-petalled female flower (called a ray flower), while each pollen-topped bump in the centre is a complete petal-less flower (called a disc flower), loaded with pollen and nectar in shallow cups that short-tongued insects can reach. Long, tubular flowers, such as foxgloves (*Digitalis*) and penstemons, are an important food source for long-tongued insects, including the garden bumblebee (*Bombus hortorum*).

HOW DO I ENCOURAGE HOUSEPLANTS TO FLOWER?

Plants invest a lot of energy in their blooms and will only flower if a sophisticated suite of sensors and biological programmes tell them that conditions are just right. For houseplants to deliver a colourful display, you must meet all these needs.

To have any hope of flowering, a plant needs to be healthy. Houseplants rely on you to supply their every need and you must give them the right light, temperature, and humidity, site them away from drafts (see pp.100–101 and pp.120–121), and keep them hydrated but not overwatered (see pp.112–115). They also need regular feeding during the growing season and repotting if they outgrow their container (see pp.110–111). Deadhead plants that have bloomed (see p.148–149) to stop energy and nutrients being wasted on making seeds.

FLOWERING CAPABILITY

Ferns do not flower and other plants, such as *Aspidistra elatior* (cast iron plant), rarely experience the tropical conditions that they need to flower inside a house. Other indoor plants, such as *Dracaena* (mother-in-law's tongue) and many cacti, will only flower when mature, and use an internal "ageing pathway" to assess whether they are old enough. Initially high levels of an ageing molecule (called microRNA156) steadily fall with each passing season, releasing the plant to flower when below a critical level.

DAILY RHYTHMS

Like most living things, plants have a "circadian clock" (body clock) which tracks the rising and setting of the sun using molecular light sensors (photoreceptors) in their leaves. Plants track the seasons by measuring the hours of darkness each night, using a molecular timer that is set running each sunset. This allows them to trigger flowering only when the chances of pollination are highest.

"Short-day" plants, such as Christmas cactus (*Schlumbergera*), will only flower in autumn and winter, when the darkness timer runs for at least 13 hours. "Long-day" plants, such as cape primrose (*Streptocarpus*), flower exclusively in brighter months when there are fewer than 10 hours of darkness (see left). Other plants, such as *Impatiens*, are "day neutral" and do not use day length as a flowering cue, so can flower from spring to autumn in favourable conditions.

WARMTH AND WATERING

Temperature can also be have a role in flowering. Many tropical plants will only bloom in consistent warmth, but *Citrus* trees and *Gardenia*, by contrast,

Daylength can influence flowering
Some houseplants will only bloom when leaf light sensors count the hours of darkness that correspond to their flowering season.

KEY

▮ Short day plants flower

▯ Long day plants flower

need a cool 13°C (55°F) at night to develop flower buds. Other plants, including many cacti and succulents, need a period of winter dormancy, without growth, before they will flower.

Orchids, such as *Phalaenopsis* (moth orchid) have unparalleled flower power, blooming for 6–12 months a year. Yet they may fail to flower if they are watered incorrectly. In the wild, orchid roots wrap around tree bark, and are coated in a spongy substance called velamen, which absorbs moisture from air, as well as water and nutrients that wash from the tree. These conditions can be tricky to mimic, but flowering is more likely if plants are watered sparingly, waiting until roots are dry and then plunging them in water for 10–15 minutes before draining, to simulate a tropical downpour. Plenty of bright light is needed for flowering, which can also be triggered by moving plants somewhere a little cooler (minimum 18°C/68°F). So-called orchid "bloom booster" feeds are not needed and research shows they may even impair flowering.

FLOWERS

SHOWY BLOOMS OPEN IN THE SEASONS WHEN POLLINATING INSECTS ARE ACTIVE IN ITS NATIVE HABITAT

BUDS FORM

KEEP COOL AND DRY IN WINTER AND BUDS DEVELOP WHEN TEMPERATURES RISE AND WATERING RESUMES

Crown cactus (*Rebutia*) remain compact and readily produce spectacular flowers in spring and summer.

SHOULD I GROW MY OWN CUT FLOWERS?

We use the language of flowers to express the inexpressible. Yet for all their evocative glory, commercially produced cut flowers come with a considerable environmental cost, providing gardeners with a great incentive to grow their own.

Year-round production of perfect blooms has created an industry that's polluting and greedy for resources. Worldwide, huge quantities of water, fertilizers, and pesticides are used to deliver blemish-free blooms, and vast greenhouses are heated using energy that could supply small towns. Flowers fill fertile land that's suitable for food crops, and roses are chilled and flown in so we can send love to our Valentine.

THE BENEFITS OF HOMEGROWN

Growing flowers for cutting in your garden is a win-win: allowing you to reap the proven emotional and mental health benefits of displaying them, while helping to reduce pollution that importing flowers causes. The blooms of many garden plants are suited to cutting, and with a little planning you can grow flowers over a long season in a small space.

FLOWERS SUITED TO CUTTING

Most annual flowers are quick and cheap to grow from seed. "Cut-and-come-again" annuals, such as *Cosmos*, *Zinnia*, and marigold (*Tagetes*), respond to picking by tirelessly producing more flowers. Single-stemmed flowers, such as sunflower (*Helianthus*) and *Celosia*, create one floral flash, so are best sown every two weeks in spring for cutting in summer.

Perennials demand more growing space, but supply flowers and foliage year in, year out. Woody shrubs, including hydrangeas and roses, yield lush foliage and pretty blooms in quantity. Soft-stemmed (herbaceous) perennials send up vase-worthy flowers in every season: hellebores in winter, showstopping peonies in summer, and Michaelmas daisies (*Aster/Symphyotrichum*) for autumn.

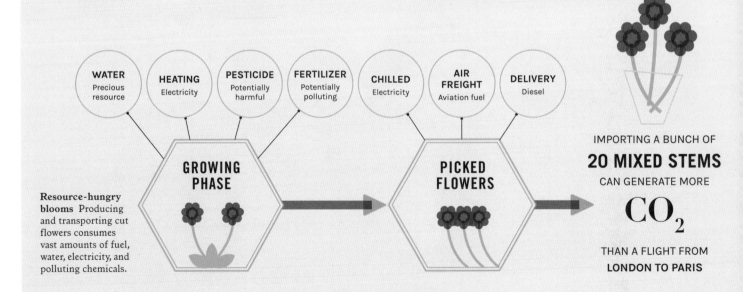

WATER Precious resource

HEATING Electricity

PESTICIDE Potentially harmful

FERTILIZER Potentially polluting

CHILLED Electricity

AIR FREIGHT Aviation fuel

DELIVERY Diesel

GROWING PHASE

PICKED FLOWERS

Resource-hungry blooms Producing and transporting cut flowers consumes vast amounts of fuel, water, electricity, and polluting chemicals.

IMPORTING A BUNCH OF
20 MIXED STEMS
CAN GENERATE MORE
CO_2
THAN A FLIGHT FROM
LONDON TO PARIS

HOW DO I EXTEND THE LIFE OF CUT FLOWERS?

Cutting a flower stem removes the life-support provided by the plant, but with some pre-arrangement care (called "conditioning"), cut flowers can be sustained for several weeks.

Without a continuous flow of sugary sap, flowers quickly wilt and enter "senescence" (programmed death), which cannot be stopped once started. Placing cut flowers in water is an immediate priority for delaying senescence. From the moment it's snipped, a cut stem will draw in air instead of water, as water already in the stem evaporates out through the leaves and flowers, in a process called transpiration (see p.22). If an air bubble (embolus) forms in the stem, all further upward flow of liquid will be blocked, condemning the flower to a short life.

DELAYING PETAL FALL

To prevent this, trim all stems by a few centimetres (up to two inches), before immediately placing them in water. Biological processes slow down at lower temperatures, so keep cut flowers cool, away from radiators and bright sunlight.

Protecting from infection can also prolong life. Clean vases thoroughly and strip all leaves and thorns which would be below the water, because they are coated with bacteria and other microbes, which will multiply in water. Cut flower food contains a drop of bleach or another disinfectant, to keep bacterial infections at bay.

Research shows that 20g (¾oz) of sugar added to a litre of vase water (1¾pints) is a suitable sap substitute for most flowers. Upward-flowing sap is naturally acidic, and a tiny dab of citric acid (0.5 grams in a litre) speeds water flow to blooms.

SAP FLOW

FLOWERS LAST LONGER IF SIMPLE ADDITIONS ARE MADE TO VASE WATER, SO IT RESEMBLES PLANT SAP

ANGLED CUT

TRIM STEMS AT A 45° ANGLE TO CREATE A BIGGER CUT SURFACE AND HELP THEM TAKE UP WATER

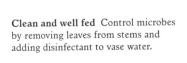

Clean and well fed Control microbes by removing leaves from stems and adding disinfectant to vase water.

WHAT IS BOLTING AND HOW DO I PREVENT IT?

A plant's ultimate purpose is to reproduce, but this drive to flower can reduce harvests of some vegetables if it happens too soon (called "bolting"). Understanding why bolting happens will help you prevent it spoiling your crops.

FLOWERING
DIVERTS SUGARS AND NUTRIENTS FROM LEAVES AND THEY BECOME SMALL AND BITTER

Spinach bolts rapidly as days lengthen in late spring, ending the crop of leaves.

When the time is right, a plant will pour all its energy and resources into the gargantuan task of producing sexual organs (flowers), then fruits and seeds. However, it's the leaves and roots of many vegetables that we are interested in, and as soon as plants begin to bolt or "go to seed" they funnel their energy into flower buds, often rendering the edible parts tough and tasteless. This can happen quite suddenly, and sometimes before plants have produced a good harvest, which is a bitter disappointment when they are otherwise healthy.

CONSIDER MATURITY AND DAY LENGTH

There are two main triggers for flowering in vegetable plants. The first is reaching the minimum size needed to support the demands of flowering and fruiting, termed "ripeness to flower". The second is the length of the dark period at night, which is how plants sense the changing seasons and is monitored using light receptors in leaves (see pp.142–143).

Many fast-growing annual vegetables, like radish, spinach, and rocket are mature enough to flower in just a few weeks and do this in late spring and early summer when days are lengthening (before 21 June in the northern hemisphere). This makes these "long-day" plants extremely prone to bolting in late spring. Experienced gardeners know to sow them very early in spring to squeeze in a harvest before flowering, and again in late summer, when they will grow without bolting in the shortening days. "Short day" plants flower through autumn and winter, when the amount of daylight is falling,

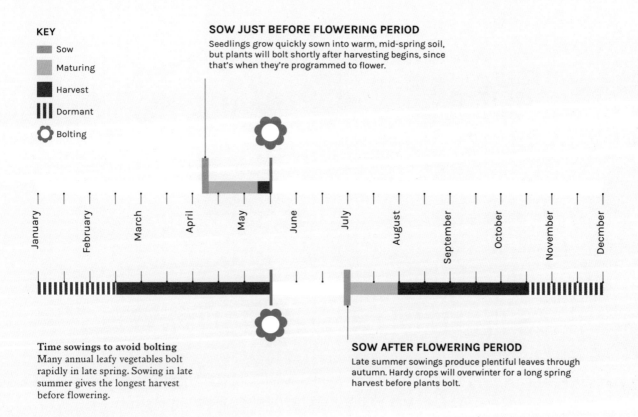

KEY

- Sow
- Maturing
- Harvest
- Dormant
- Bolting

SOW JUST BEFORE FLOWERING PERIOD
Seedlings grow quickly sown into warm, mid-spring soil, but plants will bolt shortly after harvesting begins, since that's when they're programmed to flower.

January · February · March · April · May · June · July · August · September · October · November · Decmber

Time sowings to avoid bolting
Many annual leafy vegetables bolt rapidly in late spring. Sowing in late summer gives the longest harvest before flowering.

SOW AFTER FLOWERING PERIOD
Late summer sowings produce plentiful leaves through autumn. Hardy crops will overwinter for a long spring harvest before plants bolt.

but few vegetables fall into this category and this rarely triggers bolting.

AVOID STRESSING PLANTS

Evolution has taught plants that when faced with the prospect of dying, flowering and producing seed quickly is the best way to make sure your genes live on. Stressful growing conditions can therefore initiate bolting, and extremes of temperature, or a lack of water or nutrition are commonly to blame. Mulch around plants to retain soil moisture, provide ample nutrients, and help insulate roots against temperature swings. Water regularly during hot, dry weather to avert dehydration. Avoid fertilizers formulated to promote flowering, which are high in phosphorus or potassium (rose or tomato feed) and can bring on bolting sooner.

HARVEST AT THE RIGHT TIME

Many vegetable plants are biennial, which means they spend their first growing season building up reserves to flower during the spring of their second growing season, after which they will die. Biennial vegetables, such as parsnips, carrots, leeks, and kale, will therefore usually only flower after going through the cold of winter, a process called vernalization, and sometimes only after the days have lengthened in spring. Vernalization is a complex process, similar to the way that plants break free from winter dormancy (see p.99). Avoid sowing seed outdoors too early in spring and exposing seedlings to cold, which can trigger them to bolt in their first year. These vegetables plump up their edible leaves or roots with sugars and starches in readiness for flowering in spring, but are usually harvested well before this happens.

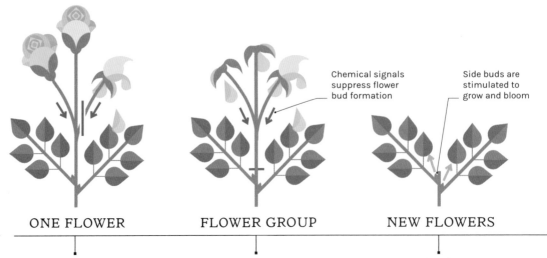

Roses

Deadheading promotes the production of new flowers.

ONE FLOWER	FLOWER GROUP	NEW FLOWERS
Snipping each flower keeps plants tidy, but chemical signals from other blooms stop lower buds growing.	**Cutting a flower stem** back above the nearest leaf, once flowers have faded, removes these chemical signals.	**Buds below the cut** are then free to grow into new shoots that will carry a fresh flush of flowers.

Chemical signals suppress flower bud formation

Side buds are stimulated to grow and bloom

SHOULD I DEADHEAD?

Every summer, secateur-wielding gardeners patrol flower beds looking for fading flowers to snip off (known as "deadheading"). While plants won't appreciate losing potential seeds, deadheading may keep them flowering for longer.

The main reason to deadhead is to coax plants into producing more flowers. A plant's goal is to make seeds, so cutting off faded flowers before seeds develop frees it from the energy-sapping burden of creating new life, diverting resources into giving flowering another go. This works well for many summer-flowering annuals, as well as some perennials and shrubs. Dahlias, for example, can flower profusely from late summer until the first frosts when regularly deadheaded.

WHY AND HOW TO DEADHEAD

Plants avoid producing too many flowers by sending out chemical signals from each bloom to suppress others forming (see p.149). If a flower is lost then this call goes silent, allowing replacement buds to develop.

PROLONG PERENNIALS
Remove spent blooms to stop seeds forming and promote further flowering.

Deadhead when flower colour starts to fade – a sign that pollination has happened, and that seeds are forming. Snip beneath the bloom, just above the first set of leaves or the next flower stem, to leave a short, tidy stump that's less prone to die-back and infection. For low-growing plants with dozens of tiny blooms, it is easier to deadhead with shears.

Another reason to deadhead is to prevent masses of unwanted seedlings springing up. For prolific self-seeders, like *Buddleja davidii*, it is responsible to deadhead to stop seeds spreading into nearby gardens and the wild.

Some seedheads are prized as features for the winter garden and as food for wildlife, including those of ornamental grasses and rose hips. Avoid deadheading these plants to allow seedheads and fruits to develop.

WHY DOES MY TREE PRODUCE FRUITS EVERY OTHER YEAR?

Fruit trees often get stuck in an on-off cycle of producing a bumper crop one year, followed by very little fruit the next. This is known as "biennial bearing" and can be smoothed out by giving your tree a little TLC.

Fluctuating fruiting rhythms are caused by a powerful plant hormone called gibberellin, which flowers and developing seeds release into the plant's sap. Gibberellin's protective job is to prevent too many flowers and fruits forming the following year.

If a tree flowers poorly one year, perhaps due to drought, spring frosts, or poor pruning, there will be less gibberellin-producing blossom and fruit. Buds that would normally be kept in check are freed to develop into flower buds for the following spring. When this profusion of flowers is pollinated, a bounty of fruit will form. High levels of gibberellin

from their seeds then restrict the next season's flowers, and so the high-low cycle continues.

CONTROL YOUR CROPS

Reduce peaks and troughs by improving the tree's growing conditions. Mulch annually with compost, water well in dry weather, and clear other vegetation for 1m (3ft) around the trunk. Probably the best way to control biennial bearing, however, is to nip off about half of the rounded flower buds during early spring in a heavy cropping year, which should reduce gibberellin and even out crops.

A chemical on and off switch
Gibberellin is a signal released from flowers and fruit that limits the amount of blossom the next spring.

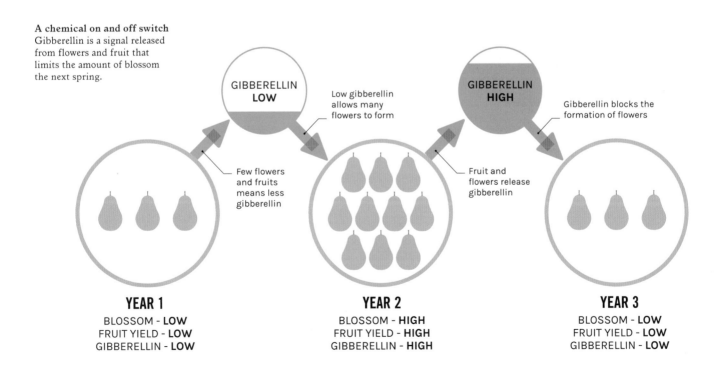

GIBBERELLIN LOW

Low gibberellin allows many flowers to form

Few flowers and fruits means less gibberellin

GIBBERELLIN HIGH

Gibberellin blocks the formation of flowers

Fruit and flowers release gibberellin

YEAR 1
BLOSSOM - **LOW**
FRUIT YIELD - **LOW**
GIBBERELLIN - **LOW**

YEAR 2
BLOSSOM - **HIGH**
FRUIT YIELD - **HIGH**
GIBBERELLIN - **HIGH**

YEAR 3
BLOSSOM - **LOW**
FRUIT YIELD - **LOW**
GIBBERELLIN - **LOW**

REBIRTH AND RENEWAL

IS MY PLANT DEAD OR JUST SLEEPING?

Every year many woody plants (trees, shrubs, and some climbers), as well as those with soft stems (herbaceous perennials) enter a kind of winter hibernation. Come early spring, it can be hard to tell if there's still life lurking in their skeletal remains.

Hibernation in animals is one of nature's best solutions for surviving harsh winters and many plants have a similar period of dormancy, when they shed leaves and eke out reserves stored during times of plenty. Plants that lose their leaves in this way are known as deciduous (meaning "fall off"). No one is certain why the switch was made from replacing leaves during the year (as evergreens do) to dumping them all in one go, but this strategy has evolved

separately many times so it's obviously effective. Like a bear putting on fat for winter, deciduous plants store their food reserves from summer somewhere safe as winter closes in. Herbaceous plants sacrifice all foliage and stems above ground and stow everything they can within their roots and underground stems, relying on their ability to regenerate from buds hidden at their base in spring. Woody plants pack energy for spring growth into the

NEW BUDS

HERBACEOUS PLANTS GROW EACH YEAR FROM BUDS THAT NESTLE AT SOIL LEVEL, IN THE PLANT'S "CROWN"

Fern New fronds remain hidden, huddled tightly beneath old leaves, until warm weather stimulates growth.

Annual cycle of dormancy and growth

A plant's internal processes control the growth of new
spring shoots and buds will not burst into life until
specific checks have been passed.

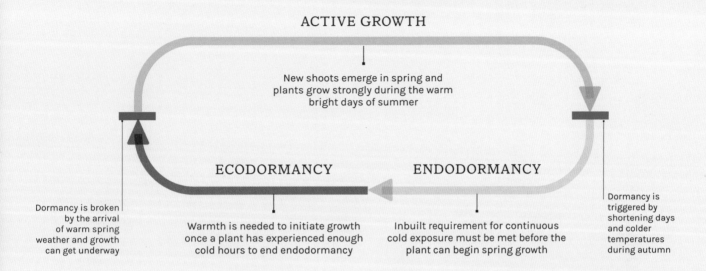

ACTIVE GROWTH

New shoots emerge in spring and
plants grow strongly during the warm
bright days of summer

ECODORMANCY   ENDODORMANCY

Dormancy is broken
by the arrival
of warm spring
weather and growth
can get underway

Warmth is needed to initiate growth
once a plant has experienced enough
cold hours to end endodormancy

Inbuilt requirement for continuous
cold exposure must be met before the
plant can begin spring growth

Dormancy is
triggered by
shortening days
and colder
temperatures
during autumn

tough tissues of their trunks and stems, as well as
their roots, and regrow in spring from the tiny buds
that cover their woody framework. Whether a plant
makes it through winter depends on that species'
inbuilt tolerance of cold (see pp.80–81), its health,
and the length and severity of the winter.

SIGNS OF NEW GROWTH

It is actually fairly easy to judge whether a tree,
shrub, or woody climber is alive or not – look for
healthy buds swelling on stems, or scrape away a
little bark, and if green is visible then the plant is
still living. Some stems may have died back while
the bulk of the plant remains healthy. By the end
of winter many herbaceous plants will be a mass of
brown stems and faded leaves, but beneath these in
early spring should be fat buds swelling at soil level.

If everything above ground looks dead and you
want to know whether there is hope of life in spring,
then dig a little around the roots. If you reveal roots
that are springy and firm then they are still alive and

kicking. Roots that are dark brown, soggy, and soft
are lifeless and rotten. If in doubt, the sure-fire way
to tell is simply to wait and see if your plant grows!

TRIGGERING SPRING GROWTH

Deciduous plants usually have a strong inbuilt drive
to power down and remain dormant until winter
has passed, and have evolved clever ways to tell the
difference between a cold autumn spell and a proper
winter. As a plant begins its winter slumber it is
flooded with growth-suppressing proteins, which act
as a chemical handbrake (termed "endodormancy").
Cold temperatures, between 0–7°C (32–45°F), cause
the gradual removal of these proteins, slowly easing
the handbrake, but a sudden warm spell snaps it
back on with a gush of growth-suppressing proteins.
When a set number of continuous cold hours has
passed (500–1,500 or more) the plant is released
from endodormancy, but is then held back by a
safety check termed "ecodormancy", which only
releases buds to swell when the temperature warms.

WHY DO LEAVES CHANGE COLOUR AND FALL IN AUTUMN?

The glorious autumn colour that precedes leaf fall in deciduous plants (see p.152) is no accident, but rather a survival strategy, carefully honed over millennia to conserve valuable resources and see them through the hardship of winter.

The resplendent golden yellows, sunset oranges, and fiery reds that herald autumn have been there all summer long – a kaleidoscope of colour hidden under a veil of green. The valuable green, light-capturing chlorophyll pigments in leaves (see pp.70–71) are constantly being broken down and replaced throughout spring and summer, but as the light fades and the days grow cold, deciduous trees and shrubs stop manufacturing them as they prepare to shed their leaves and bed down for the winter (see pp.152–153). Before casting off their fluttering solar panels, however, plants try to drain every morsel of goodness out of them, sucking out the primary nutrients phosphorous and nitrogen (see pp.122–123), valuable chlorophyll remnants, and breaking down leaf starches into sugars for safe keeping in the trunk and roots (see pp.152–153).

WHAT LIES BENEATH

Alongside chlorophyll, lower levels of "accessory" leaf pigments also lend a hand in food production. As leaves are drained of green, the burnt oranges and canary yellows of the six carotenoid pigments are unveiled. Some plants then bow out with an extra farewell flourish of scarlet and mulberry purple, as seen in sugar maple (*Acer saccharum*) and dogwood (*Cornus florida*).

Scientists aren't sure exactly why some plants pump these ruddy protective chemicals, called anthocyanins (meaning "flower-blue") into their leaves at this time: it seems these pigments are off-putting to insect predators looking to get an autumn snack, and may serve as a protective sunscreen for ageing leaves now bereft of their green light absorbers. The most breathtaking autumn colours follow cool nights, which speed up chlorophyll destruction, alongside bright, dry sunny days, which slow down sugar's final escape from leaves, accelerating the production of anthocyanins and intensifying their reddish glow.

Plants that have red or purplish leaves year-round have an abundance of anthocyanin pigments that mask their chlorophyll and turn red in autumn as the green pigment breaks down. Leaves that simply turn brown lack this colourful underlay and reveal the colour of their tannin-impregnated skeleton.

DRAINED AND DISCARDED

As well as transforming its colour, the ripples of chemical changes within a leaf that happen in autumn also cause a cuff of cells at the base of the leaf stalk to start contracting. Gradually tightening with each day as its leaf is emptied of the final dregs of goodness, this band, known as the "abscission" zone, pinches off the base of the stalk to leave an airtight scar, causing the leaf to cartwheel down to its final resting place on the soil below. Leaf shedding seems to have evolved in temperate climates because it is a safer bet to ditch large leaves in winter, when they are liable to be damaged or ripped off by wind and snow, especially given that they won't produce much food during the short days anyway. Shedding leaves also prevents precious water escaping from leaf pores by transpiration (see p.22).

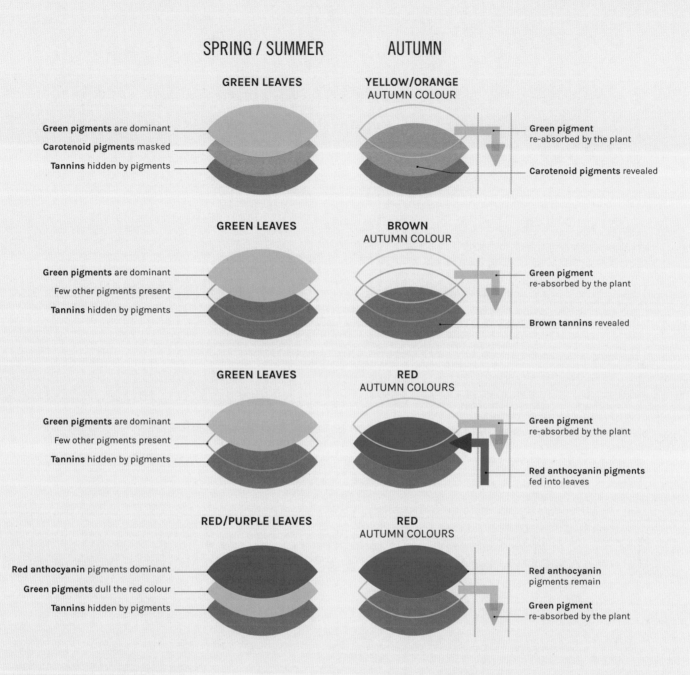

SPRING / SUMMER   AUTUMN

GREEN LEAVES

YELLOW/ORANGE
AUTUMN COLOUR

Green pigments are dominant

Carotenoid pigments masked

Tannins hidden by pigments

Green pigment
re-absorbed by the plant

Carotenoid pigments revealed

GREEN LEAVES

BROWN
AUTUMN COLOUR

Green pigments are dominant

Few other pigments present

Tannins hidden by pigments

Green pigment
re-absorbed by the plant

Brown tannins revealed

GREEN LEAVES

RED
AUTUMN COLOURS

Green pigments are dominant

Few other pigments present

Tannins hidden by pigments

Green pigment
re-absorbed by the plant

Red anthocyanin pigments
fed into leaves

RED/PURPLE LEAVES

RED
AUTUMN COLOURS

Red anthocyanin pigments dominant

Green pigments dull the red colour

Tannins hidden by pigments

Red anthocyanin
pigments remain

Green pigment
re-absorbed by the plant

Colour changes in autumn leaves (left)
The transformation that leaves undergo as
the weather cools and the days shorten is
dictated by the pigments present as plants
re-absorb their green chlorophyll.

Inherited timing

DECIDUOUS PLANTS ARE PROGRAMMED FOR ANNUAL WINTER DORMANCY.
Plants learn the seasonal patterns for their region and pass this learning on
to their offspring through chemical messages tagged onto their seedlings'
DNA (a process called epigenetics). If a tree's seeds are grown many miles
away from the parent, then the sapling will drop its leaves according to the
daylength in their parent's location, regardless of where they now are.

SHOULD I KNOT THE FOLIAGE OF SPENT BULB PLANTS?

Spring-flowering bulbs are a ray of sunshine after winter, but their blooms are followed by the drawn-out withering of their dying leaves. This can upset gardeners who prefer neatness, but forcing nature to be tidy is not usually for the best.

A time-honoured tradition is to neatly knot or braid the fading foliage of larger-leaved bulbs, such as daffodils. Some gardeners may even snip leaves off before they look tatty. Mutilating the leaves in this way, however, limits the plant's ability to refuel its bulb after the effort of flowering.

LET LEAVES DO THEIR JOB

While they remain green, leaves are busy photosynthesizing (see pp.70–71) and sending sugar made by this process down to the bulb for safe keeping. Here the sugars are chemically welded together into starch and bundled into "granules", which can be stored, ready to be turned back into sugar when needed. The more starch granules that can be stuffed into each cell within a bulb, the stronger next spring's growth will be.

Knotting leaves reduces the leaf area exposed to sunlight, and therefore their ability to make sugar. Worse still would be to cut the leaves off before they have finished their feeding role. If leaves are allowed to die back naturally, or are left for at least six weeks after flowering, then bulbs should go on blooming for many years. Plus, even more sugary goodness will be piped down into the bulb if faded flowers are cut off ("deadheaded") before seeds form (see pp.148–149).

If you wish to disguise the ageing leaves, try planting them among other plants, such as herbaceous perennials, that will produce a flush of new foliage just as the bulb leaves are doing their final job.

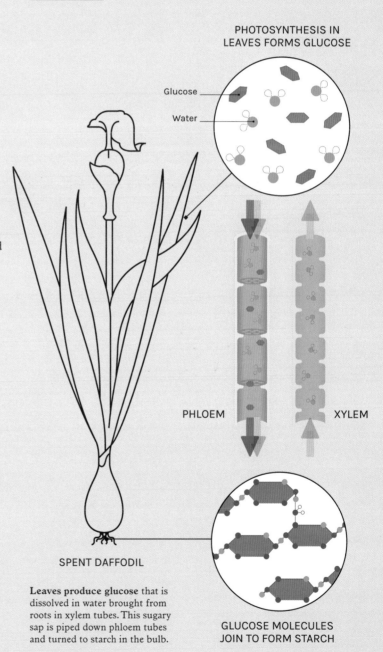

PHOTOSYNTHESIS IN LEAVES FORMS GLUCOSE

Glucose

Water

PHLOEM

XYLEM

SPENT DAFFODIL

GLUCOSE MOLECULES JOIN TO FORM STARCH

Leaves produce glucose that is dissolved in water brought from roots in xylem tubes. This sugary sap is piped down phloem tubes and turned to starch in the bulb.

SHOULD I LIFT AND STORE BULBS AFTER FLOWERING?

Many plants have evolved underground storage organs (see pp.96–97) to survive extreme growing conditions. Why then do some gardeners dig up ("lift") bulbs after flowering to stow them in a paper bag or cardboard box?

Newly purchased bulbs have been grown in optimal conditions and are bursting with energy, ready to flower. Once planted they can usually be left in place to bloom year after year, but sometimes they are best lifted, when evolution has not prepared them to thrive in your garden's climate.

PROTECTION FOR TENDER BULBS

Dahlias, for example, evolved in mountainous Mexico and, along with other tender (see pp.80–81) plants with underground storage organs, such as *Gladiolus* and *Canna*, are not equipped to deal with damp, sub-zero winters. Faced with the risk of frost damage and rot (a fungal infection), it can be safer to lift faded plants in autumn and, once clean and dry,

store their bulbs somewhere cool and frost-free for winter. Alternatively, a thick mulch gives dormant plants some frost protection (see pp.160–161).

SIMULATE A BAKING SUMMER

Tulips are cold hardy, but originally hail from central Asia, where summers are hot and dry. In cool, wet summers their bulbs won't get the right moisture and temperature signals to stay dormant and may rot or flower poorly the following year. For this reason, tulips can be lifted when their leaves have died back, and their bulbs are packed with starches (see opposite). They can then see out their dormancy stored in nets or paper bags in a dry place, at 18–20°C (64–68°F), before being replanted in autumn.

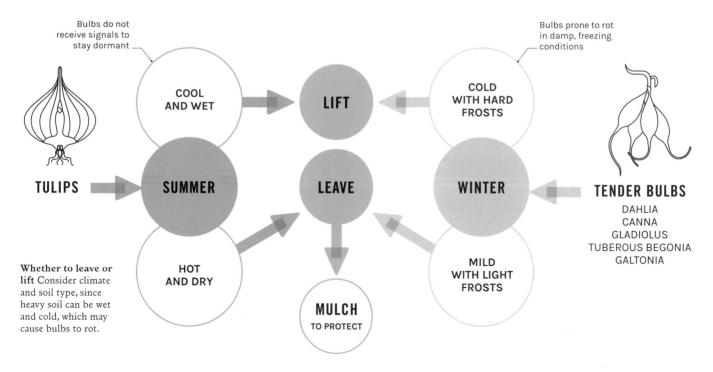

Bulbs do not receive signals to stay dormant

Bulbs prone to rot in damp, freezing conditions

COOL AND WET

LIFT

COLD WITH HARD FROSTS

TULIPS

SUMMER

LEAVE

WINTER

TENDER BULBS
DAHLIA
CANNA
GLADIOLUS
TUBEROUS BEGONIA
GALTONIA

HOT AND DRY

MILD WITH LIGHT FROSTS

MULCH
TO PROTECT

Whether to leave or lift Consider climate and soil type, since heavy soil can be wet and cold, which may cause bulbs to rot.

HOW DO I CARE FOR MY GARDEN OVER WINTER?

A winter garden can seem a wasteland, where fallen leaves turn slimy in the rain, trees and shrubs are reduced to skeletal silhouettes, and brittle seedheads replace elegant blooms. But there is still life, and much can be done to care for it.

As sunlight and temperatures decline, all of life's processes slow – root growth becomes sluggish, photosynthesis (see pp.70–71) slumps, and when soil cools below 10°C (50°F), the bustling bacteria of the soil food web (see pp.44–45) grind to a standstill.

SOW, PLANT, AND PRUNE FOR SPRING

Although little is growing, seeds and bulbs that need a winter chill (see pp.76–77 and p.99) can be planted. Similarly, deciduous trees and shrubs are best planted in late autumn or winter to give roots plenty of time to establish, so that come spring the root hairs (see pp.104–105) are ready to suck up

nutrients and water for an explosion of growth. This winter dormancy (see pp.152–153) is also the ideal time to prune many woody plants (see pp.168–169). For plants that are less hardy to cold conditions (see pp.80–81), a thick mulch of compost around their base in autumn will protect them through the year's darkest days (see pp.160–161).

DON'T RUSH TO CLEAR AND CUT

Traditional advice in many gardening books and websites will tell you that winter is the time to tidy up the "mess" of dead plant matter and to dig over bare patches of earth to "aerate" soil. Science now

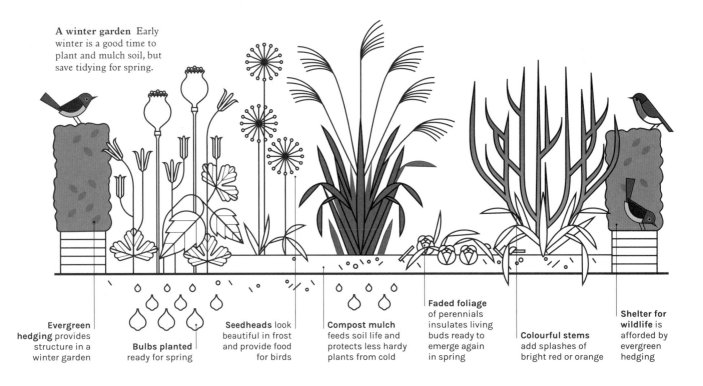

A winter garden Early winter is a good time to plant and mulch soil, but save tidying for spring.

Evergreen hedging provides structure in a winter garden

Bulbs planted ready for spring

Seedheads look beautiful in frost and provide food for birds

Compost mulch feeds soil life and protects less hardy plants from cold

Faded foliage of perennials insulates living buds ready to emerge again in spring

Colourful stems add splashes of bright red or orange

Shelter for wildlife is afforded by evergreen hedging

says that these chores should be consigned to the history books, for while they may make you feel good, they leave the winter garden bare and don't help your soil, plants, or garden wildlife.

Cutting back the faded growth of deciduous herbaceous perennials, such as *Crocosmia*, *Rudbeckia*, and *Nepeta* (catmint), to just above the dormant base has conventionally been a winter task to keep a patch looking tidy and sometimes to prevent seeds falling and germinating. But by prizing neatness, you remove flowerheads and seedheads that add structure and interest during the bleakest months, and rob resident birds and insects of a useful habitat and food supply – it is far better to leave cutting back until spring. Leaving the previous season's stems and foliage also offers protection to the buds and roots of less hardy plants, such as penstemons, which may succumb if trimmed back too early, and helps to cover soil and protect it from erosion and damage during harsh winter weather.

Fallen leaves should also be appreciated as part of the cycle of life. Left on soil or at the base of hedges they will replenish the soil with nutrients as they decompose and be fuel for the soil food web. They should be raked off lawns or plants to prevent them smothering growth, but are a valuable resource to collect to add to the compost heap or a separate heap to make "leafmould" (see pp.190–191).

NURTURE YOUR SOIL
Digging over soil in winter may be good exercise, but with each downward slice of your spade you will be severing plant-nourishing fungal threads and collapsing tiny tunnels bored out by worms that allow water and

air to drain. Lifting and turning over each clod of earth exposes buried carbon-chomping bacteria to the open air, kick starting them into digesting the soil's organic matter and releasing puffs of invisible carbon dioxide. You may feel that you are adding air to the soil (aerating) by digging it over, but this action destroys its structure and encourages nutrients to wash away. It's far better to apply a layer of compost mulch in autumn and let the soil organisms do the hard work (see pp.42–43).

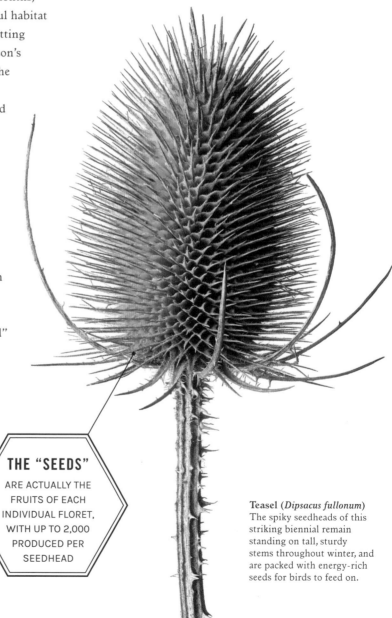

THE "SEEDS" ARE ACTUALLY THE FRUITS OF EACH INDIVIDUAL FLORET, WITH UP TO 2,000 PRODUCED PER SEEDHEAD

Teasel (*Dipsacus fullonum*) The spiky seedheads of this striking biennial remain standing on tall, sturdy stems throughout winter, and are packed with energy-rich seeds for birds to feed on.

WHAT OFFERS THE BEST PROTECTION AGAINST FREEZING TEMPERATURES?

Imagine for a moment that you're outside, winter is on its way, but your feet are glued to the ground. This is the reality for plants in temperate regions, where a little extra protection can make this annual ordeal less damaging and easier to endure.

Plants that are not fully hardy in your region's climate (see pp.80–81) may perish in winter if planted in the wrong position and without some extra protection. Avoid planting all but the hardiest plants in dips or low-lying areas, where colder, denser air will gather, causing a "frost pocket". Locating less hardy plants in sunny, south-facing areas will give them the best chance of coming through the winter (see pp.36–37).

A WARMING BLANKET

Fabric crop covers can help to protect your garden plants from the cold, and are one of the most widely used defences against sub-zero temperatures. They are especially useful in early spring, to protect new growth and newly planted seedlings. We warm-blooded animals constantly pump out heat, which our clothing captures. Plants generate virtually no heat, so crop covers can only hold on to daytime warmth created by the sun beating down on the soil. At night, stored heat radiates skywards, potentially causing frost (see pp.118–119), but a crop cover retains some of this heat, keeping the plants beneath warmer for longer.

Light, permeable fabrics, like horticultural fleece, are usually the best option if you plan to leave the cover in place, because they allow light, rain, and air through to keep plants in good health and won't flatten them if they get wet. Any fabric that doesn't let light through, like opaque plastic, will need to be lifted during the day, and will also need to be held clear of plants on a frame to avoid their weight

Covers for crops and soil
Protect young plants from frosts in spring using a fabric, plastic, or glass crop cover. A compost mulch insulates established plants.

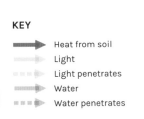

KEY

→	Heat from soil
→	Light
⤏	Light penetrates
▪▪▪▶	Water
▪▪▪▶	Water penetrates

FLEECE

Plants thrive under this light, permeable cover in cold weather and it's easy to add if frost is forecast.

PLASTIC TUNNEL

Clear plastic retains heat from soil and sunlight, but plants need watering and ventilation to stay healthy.

COLDFRAME

A glass or plastic lid keeps plants insulated and allows in light, but is propped open for daytime ventilation.

causing damage. An insulating layer of dry straw can be packed around tender plants, like bananas or tree ferns, held in place with wire mesh, and kept dry through winter with a outer cover of polythene sheet. Such covers must be removed as soon as the weather warms sufficiently in late spring.

SOLID STRUCTURES

Clear plastic cloches, polytunnels, low coldframes, and greenhouses also trap the sun's warming rays through the day to raise daytime temperatures, but because they are not insulated, these unheated structures do little more than crop covers to prevent temperatures dipping to damaging lows at night. They exclude rain, which can be an advantage for dormant plants that need dry winter conditions, but any plants in growth will need to be watered. It's also vital to provide ventilation by opening lids, doors, or vents during the day, to stop temperatures climbing too high on sunny days and prevent high air humidity, which will leave plants prone to fungal infections and rots.

Winter winds can be damaging for exposed plants, blowing away precious moisture and potentially drying leaves to a crisp (see pp.118–119). Crop covers offer useful protection against wind, although they can be blown away themselves, while greenhouses and coldframes are a formidable defence if a plant can be moved undercover. A useful alternative in exposed gardens is to shield less hardy plants, or even the whole garden, by planting a "shelter belt" of hardy shrubs or trees on the side of the prevailing wind (see pp.36–37).

MULCHING IS A MUST

As the air temperature fluctuates with the rising and setting of the sun, soil lags behind because it doesn't give up its heat so easily. Damp clay type soils hold onto daytime heat better than lighter sandy soils, but science shows that regardless of your soil type, putting an insulating layer of compost, woodchip, or straw mulch around plants in late autumn allows soil to hold on to more of its heat and helps prevent roots becoming frozen. This can make the difference between success and failure with plants that are borderline hardy, such as dahlias or *Agapanthus*. These organic mulches are also extremely beneficial to soil health (see pp.42–43).

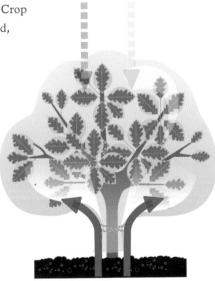

Protection for larger plants
Mature plants that can't be moved undercover can be shielded from cold in situ. Covers need to be in place before the first frost and removed before spring growth begins.

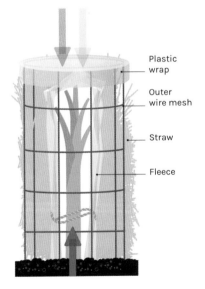

Plastic wrap
Outer wire mesh
Straw
Fleece

MULCH

A thick layer of compost or bark chips holds residual heat from soil, protecting roots and dormant buds.

FLEECE WRAP

Swathing borderline hardy plants in fleece provides useful protection from icy winds. Light can also penetrate, which is essential for evergreens.

DRY STRAW

Thick, dry insulation is needed to protect tender plants. Pack straw around the plant, secure with mesh, and keep dry with plastic wrapping.

WHY DO SOME VEGETABLES TASTE BETTER AFTER FROST?

Not only are some winter vegetables built to survive the cold in perfect condition, but they also taste better once the temperature in the garden tumbles – Brussels sprouts really are sweeter if you save them for Christmas.

Termed "hardy" for their ability to withstand sub-zero temperatures, Brussels sprouts, kale, parsnips, and other vegetables all have a defensive system that kicks in when the temperature drops below 4°C (39°F). Anticipating a frost, sensors within plants detect falling temperatures and start digesting stored starches into sugars. These act as a natural antifreeze in the watery plant cells, preventing them being ripped apart by expanding ice crystals (see p.118–119).

COLD-INDUCED SWEETENING
Sugary water freezes at a lower temperature than pure water, preventing ice crystals forming inside a plant, in the same way as salt spread on roads stops ice forming (sugar would do the same thing, but this might cause animals to put themselves in harm's way while licking the tarmac).

While these vegetables are perfectly good when harvested in autumn, holding off picking and lifting them until frost has set in will mean they have a noticeably sweeter flavour, which most people prefer. Those with lots of starch to begin with are especially delicious with cold-induced sweetening, such as parsnips, swede, and turnips. Less starchy vegetables such as leeks, Brussels sprouts, kale, and spinach also get a sweetness boost. Even sweet potatoes get sweeter after a little frost, providing this "tender" crop has finished growing and is well protected beneath the soil.

Potatoes are the important exception that should always be dug up before there is any risk of frost. Extra sugar caused by exposure to cold will make them brown much faster when fried, grilled, or roasted, so that they look burnt when cooked.

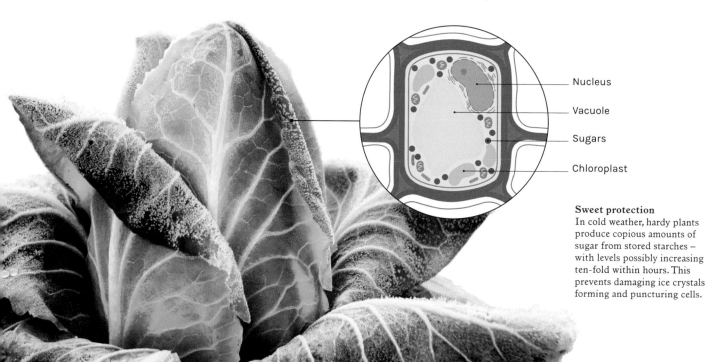

Nucleus

Vacuole

Sugars

Chloroplast

Sweet protection
In cold weather, hardy plants produce copious amounts of sugar from stored starches – with levels possibly increasing ten-fold within hours. This prevents damaging ice crystals forming and puncturing cells.

WHY DO MY HOUSEPLANTS DIE OVER WINTER?

Houseplants may be sheltered from the weather, but even indoors there is no hiding from the seasons. Low light levels, cold draughts, central heating, and incorrect care all cause problems during a period when many plants put on little growth.

WATERING

Short days mean less solar power is available for photosynthesis and, as food-making is cranked down, the breathing pores (stomata) on the underside of the leaves spend more time closed, reducing water loss through transpiration (see p.22). All this reduces the need for roots to take up water and so soil will stay damp for longer, so reduce watering accordingly, while remembering that you may need to increase humidity around leaves in the dry air of a centrally heated home (see pp.120–121). Overwatering encourages rotting, but occasional watering with tepid water mimics the plant's natural environment and helps keep roots healthy.

LIGHT AND WARMTH

Even though many houseplants evolved in the shady understory of a forest, they can struggle to get enough sunlight during winter. They are usually best kept in a bright position, but out of direct sunlight, which can damage delicate leaves. Rotate plants frequently, because shaded stems may brown or become elongated and "leggy" as they curve and stretch towards the light (see p.12).

Most houseplants hail from the tropics and can suffer "chilling injury" at temperatures below 10–15°C (50–59°F). When plants are too cool roots struggle to take up nutrients, and life processes grind to a halt as the protective barrier around plant cells becomes brittle and leaky, while the liquid inside turns thick and gloopy. All of which results in leaves losing their colour and wilting. Prevent this by keeping plants in a warm room – not above a radiator – and away from cold draughts. Always avoid shutting them behind curtains, where conditions can be cold at night.

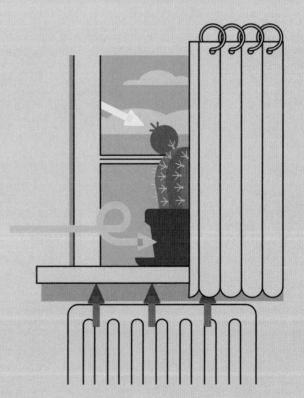

Windowsills can be an unsuitable winter home:
A sunny windowsill can be fraught with problems for houseplants in winter. Intense sunlight and cold draughts dehydrate and damage leaves, along with warm, dry air from a radiator below. Temperatures also plummet between curtains and a window at night.

KEY

➡ Dry air
➡ Intense sunlight
➡ Cold draught

WHAT IS THE AIM OF PRUNING?

To sever limbs from plants can seem like a cruel mutilation, after all, no plant needs human intervention to grow, flower, or fruit. But careful pruning can be essential to keep plants healthy, manipulate their size and shape, and promote plentiful blooms.

ALWAYS REMOVE

DEAD
LOOK FOR BROWN, DRY BRANCHES WITH SHRIVELLED BARK AND NO HEALTHY BUDS.

DISEASED
SYMPTOMS INCLUDE: ROUGH GROWTHS ON BRANCHES, WEEPING OR OOZING FROM BARK, FUNGAL GROWTH.

DAMAGED
ANY INJURY TO BARK FROM BRANCHES RUBBING, BREAKING IN WIND, OR ANIMALS FEEDING.

The first priority when pruning is always to prevent and remove disease. Where there is evidence of disease or damage, plants benefit from surgery to keep them well and free from infection. Like humans, a plant's outer "skin" is its first line of defence against infection. Branches that are dry and dead, and those with damaged bark, are an open door for would-be invaders and need to be removed, or pruned back promptly to just above a healthy bud. Branches that cross will rub together, damaging their bark, so should also be removed. Densely packed ("congested") branches and leaves in the centre of a shrub or tree prevent free air circulation and create humid conditions that are a breeding ground for disease-causing fungi. Pruning to "thin out" congested branches (see pp.170–171) improves airflow and allows light to reach the plant's centre, making it harder for harmful fungi to get a foothold.

SHAPING AND TRAINING

When it comes to caring for trees, shrubs, and woody climbers, it is easiest for you (and less traumatic for the plant) to establish the desired branch structure by pruning a young plant before branches develop and thicken ("formative pruning"). Tying branches to supports to direct their growth ("training") should also begin while they are young and supple. Pruning can

Basic pruning for health keeps plants strong and disease-free. Thinning cuts remove congested growth and mean that resources from the root system are split between a reduced number of branches, resulting in more vigorous growth.

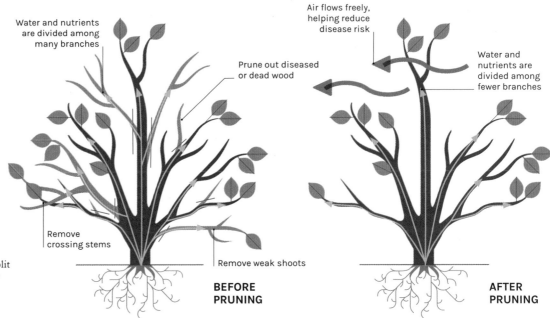

Water and nutrients are divided among many branches

Prune out diseased or dead wood

Remove crossing stems

Remove weak shoots

BEFORE PRUNING

Air flows freely, helping reduce disease risk

Water and nutrients are divided among fewer branches

AFTER PRUNING

Creating a vase-shaped fruit tree

Buying a mature tree that's pruned to shape is expensive. Instead consider buying a young tree and carrying out formative pruning yourself. An open vase shape is an attractive and popular option for garden trees.

Improved air circulation afforded by vase shape helps to prevent the formation of disease-causing fungi, bacteria, and moulds

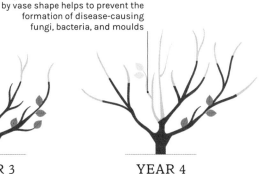

YEAR 1	YEAR 2	YEAR 3	YEAR 4
Plant a two-year-old tree between late autumn and early spring. Cut the leading stem back to just above a bud or branch, leaving the desired length of trunk.	**Select three branches** growing at wide angles to the trunk and cut back by about half to outward-facing buds (see p.171). Remove unwanted branches.	**Prune the leading shoots** of each branch back by about half of the previous year's growth to promote further branching. Remove weak shoots.	**The branch framework** is formed, but can continue to be shaped. Cut back any weak shoots or those growing into the centre of the tree.

control a plant's shape, size, and structure. Many trees are strongest with a single trunk (termed a "central leader"). If the main growing point is damaged or two branches compete for the top spot, trees can develop "double leaders" which will be pulled apart by increasingly heavy side branches and may eventually cause the tree to split down the middle. Avoid this by pruning off any vigorous side branch that threatens to compete with or overtake the leading stem.

Sometimes fruit trees are pruned to produce a vase-shaped pattern of branches with an "open centre", where three or more main branches grow out evenly from a central trunk. This structure allows air and light into the canopy to maintain healthy growth, help ripen fruit, and make picking easier, and is created by removing the central leader of a young tree to stimulate side branches to form (see above).

PROMOTE YOUNG GROWTH

Sometimes old, unproductive branches hold a plant back, placing a drain on precious resources, and are often best pruned right back to the base if they become bare or fail to flower. This diverts energy to

new growth and prevents branches becoming tangled and congested. Once established, shrubs that form clusters of upright stems, such as Japanese rose (*Kerria japonica*), mock orange (*Philadelphus*), and blackcurrants (*Ribes nigrum*), benefit from an annual culling of around one quarter of the branches over three or four years old. Many old, overgrown shrubs can even be given a new lease of life by pruning them back close to the base to stimulate new growth.

Naturally vigorous shrubs and climbers that flower in summer, like bush and hybrid tea roses, butterfly bush (*Buddleja davidii*), and late summer-flowering *Clematis*, are so energetic that pruning them back to a stumpy framework of branches early each year will result in more compact growth and a more glorious display of larger flowers. Other shrubs can be pruned hard early in the year to maintain their youthful good looks, rather than promote flowering. These include common dogwood (*Cornus sanguinea*) which has stunning, fiery orange bark on young stems, along with elder (*Sambucus*) and smoke bush (*Cotinus coggygria*), which both develop larger, lusher leaves on new growth.

WHAT HAPPENS AFTER I MAKE A PRUNING CUT?

Limb loss in the plant kingdom is not a death sentence, as it can be in animals. Plants heal effectively, can actually be rejuvenated by pruning (see pp.164–165), and regenerate in a predictable way, which allows you to judge the best places to cut.

After a pruning cut, new youthful growth erupts from side buds lower down the stem. Plants are only able to regenerate from buds, where the embryonic "meristem" cells that divide to produce new shoots are neatly bundled (see pp.136–137). Plant growth is usually channelled into the tip – or "apex" – of upward pointing shoots. This helps boost food production by allowing stems to lengthen towards sunlight and so harvest more of its life-giving energy in their leaves.

APICAL DOMINANCE

To ensure that side branches do not divert resources from this upward growth, the shoot apex chemically subdues their development by releasing a substance called "auxin", which is known as a plant hormone. Auxin seeps down the stem from the growing point, suppressing any side bud it meets, to create an effect known as "apical dominance". This shapes the form of plants both during their natural growth and after pruning.

Cut off the growing point, however, and the sedating flow of auxin instantly ceases, allowing sugars previously reserved for apical growth to flood into dormant side buds. Those buds closest to the amputated tip were the most heavily suppressed and so will respond by growing more vigorously. This creates more branching and bushier growth, which

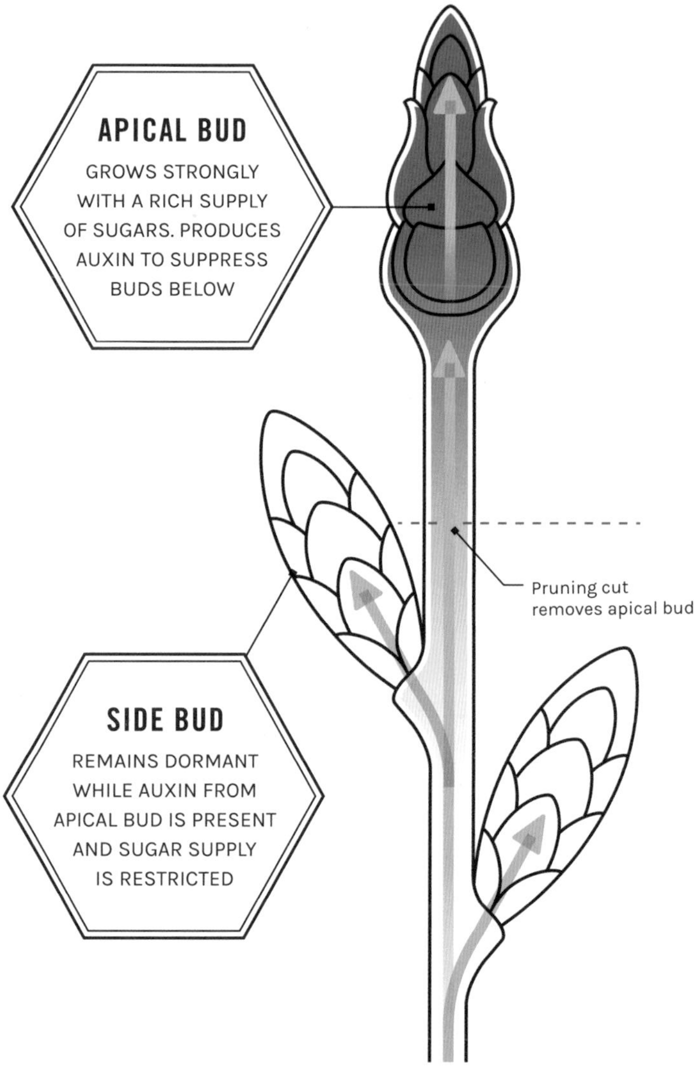
APICAL BUD
GROWS STRONGLY WITH A RICH SUPPLY OF SUGARS. PRODUCES AUXIN TO SUPPRESS BUDS BELOW
Pruning cut removes apical bud
SIDE BUD
REMAINS DORMANT WHILE AUXIN FROM APICAL BUD IS PRESENT AND SUGAR SUPPLY IS RESTRICTED

APICAL BUD

Branches lengthen via apical buds at their tip, and their dominance encourages tall growth to reach towards the light.

Apical dominance

In woody plants, growth is concentrated in the tip – or "apex" – of a branch, to the extent that side buds lower down are inhibited from growing due to the flow of the hormone auxin, in a process called "apical dominance". Pruning disrupts this dominance.

improves the appearance and flowering potential of many garden plants. In trees, one of these new branches may grow upwards to take the helm as the dominant apical bud, known as the "leading shoot". New growth after pruning is also supercharged, thanks to the surging supply of water and nutrients from an established root system.

HEALING

After any kind of cut, a plant can suffer the same maladies as we do – bleeding and infection. Because plants have no beating heart, sap loss is rarely life threatening, but infection is a real peril. We have immune cells that hurtle through blood to mend and defend a wound, but in plants, every cell must protect and repel.

Within minutes of a blade slicing through a branch, the cut surface produces first a chemical and then a physical seal. Tough-as-wood lignin hardens the surface, while a waxy substance called suberin forms a waterproof scab (called a "callus"). But these wounds do not heal like ours: damaged plant parts cannot be repaired and so are quarantined from living tissue with an internal protective barrier, in a process known as "compartmentalization".

Any stump of tissue that is left above a healthy, growing bud will quickly be compartmentalized and gradually die back as their supply of water and nutrients is cut off. The longer this stump, the greater the risk of it becoming infected with fungi or bacteria, so pruning cuts (see pp.170–171) should always be made as close to a healthy bud as possible.

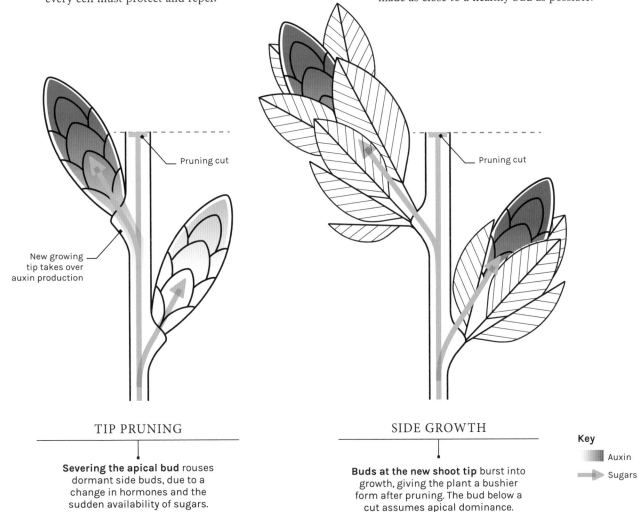

Pruning cut

New growing tip takes over auxin production

Pruning cut

TIP PRUNING

Severing the apical bud rouses dormant side buds, due to a change in hormones and the sudden availability of sugars.

SIDE GROWTH

Buds at the new shoot tip burst into growth, giving the plant a bushier form after pruning. The bud below a cut assumes apical dominance.

Key

Auxin

Sugars

IS THERE A CORRECT TIME OF YEAR TO PRUNE?

Much is written and said about the "right" time to prune different plants, since this is often important to improve flowering and fruiting, and prevent frost damage. But most plants are resilient enough to recover if accidentally pruned at the "wrong" time of year.

There is no golden rule for pruning all plants. It helps to remember that plants generally rest in winter and do most of their growing in spring and summer, but it is the particulars of each species's growing and flowering "habit" that influence when is the best time to prune.

PRUNE HARD DURING DORMANCY

As a general rule, winter is a good time to carry out any major pruning on deciduous trees and shrubs, including renovation, because they then have a full growing season to recover and regrow. During winter, these plants are dormant: all but the most

essential life functions are powered down, and all of the sugars and nutrients from its leaves have been stored in the roots and trunk. In this leafless, ungrowing state the structure of branches and buds is clearly visible, helping you to make clean, well-considered cuts. A dormant plant will have plentiful energy to produce new growth come spring. For this reason, pruning in winter is said to "stimulate" growth, while pruning to remove leafy branches in summer removes some of a plant's energy source and "restricts" growth (see pp.172–173).

In dormancy, however, the plant can do little healing, so it's best to hold off pruning until the second half of winter, when there is only a short wait until spring's restorative arrival. Open wounds are particularly vulnerable to frost for 10 days after pruning, so if a cold snap is forecast then put away the secateurs until warmer weather arrives. Delay pruning evergreens until well into spring, when the risk of frost has completely passed, because their new shoots are tender and very easily damaged.

There are also exceptions among deciduous plants, where winter pruning can be damaging.

KEY

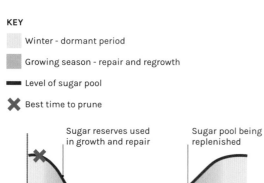

- Winter - dormant period
- Growing season - repair and regrowth
- ▬ Level of sugar pool
- ✖ Best time to prune

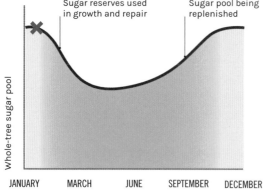

Sugar reserves used in growth and repair

Sugar pool being replenished

Whole-tree sugar pool

JANUARY   MARCH   JUNE   SEPTEMBER   DECEMBER

When to prune during dormancy Late winter is the best time to prune most deciduous woody plants, when they are packed with sugar reserves to power new growth and can quickly heal cuts when they launch into spring growth.

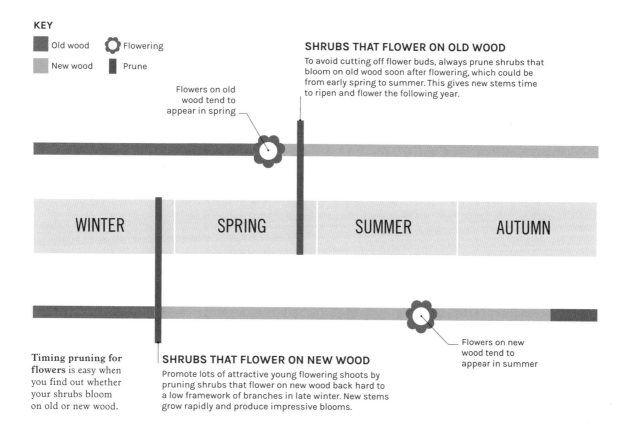

KEY

- ■ Old wood
- ■ New wood
- ✿ Flowering
- ▮ Prune

Flowers on old wood tend to appear in spring

SHRUBS THAT FLOWER ON OLD WOOD

To avoid cutting off flower buds, always prune shrubs that bloom on old wood soon after flowering, which could be from early spring to summer. This gives new stems time to ripen and flower the following year.

| WINTER | SPRING | SUMMER | AUTUMN |

Timing pruning for flowers is easy when you find out whether your shrubs bloom on old or new wood.

SHRUBS THAT FLOWER ON NEW WOOD

Promote lots of attractive young flowering shoots by pruning shrubs that flower on new wood back hard to a low framework of branches in late winter. New stems grow rapidly and produce impressive blooms.

Flowers on new wood tend to appear in summer

For example, trees that produce stone fruits (*Prunus*) are vulnerable to specific infections (bacterial canker and silver leaf disease) in winter and are best pruned in summer. While some woody plants, including walnut trees (*Juglans*), birches (*Betula*), maples (*Acer*), and figs (*Ficus*), start piping their sugary sap up from their roots very early and will "bleed" profusely if pruned too late in winter.

PRUNING TO BOOST BLOOMS

When pruning shrubs and climbers grown for their floral display, it is helpful to know whether each year's flowers grow on "new" or "old" wood. Some shrubs, such as *Buddleja davidii*, hardy *Fuchsia*, *Hibiscus*, *Hydrangea paniculata*, and *Spiraea japonica*, flower from midsummer on fresh shoots that have grown that spring (new wood). These are best pruned hard in late winter, so that their new shoots have time to grow, mature, and flower.

Others (*Chaenomeles*, *Forsythia*, *Syringa*/lilac, *Weigela*) produce flowers only on previous years' growth (old wood). These shrubs typically flower earlier in the year and are best pruned after their blooms have faded to allow new growth time to mature so that it can produce flowers the following year. This also removes the risk of pruning off the old wood that is due to bloom, and inadvertently wiping out that year's display. If in doubt, don't prune any shrub that flowers before June during winter or before flowering.

DOES IT MATTER WHERE I MAKE A PRUNING CUT?

Like surgeons, all gardeners must know where and how to cut when pruning. This sounds complicated, but there are basically two options: to amputate an entire branch or stem at its base, or to remove a portion of a branch or stem by cutting at a point along its length.

Before you begin, remember that it takes a lot of energy for a plant to recover from an injury. Many woody plants can handle heavy pruning, but if in doubt remove no more than one fifth of the branches at one time. Start with clear goals and regularly take a step back to get an overall view of your progress. It's safer to cut less when you begin, because you can always come back to prune more if necessary.

Like injuries to our skin, pruning cuts are an open doorway for infection, and it is important to use sharp tools to make clean cuts that will heal quickly. It is a good idea to clean and sterilize blades too, although there is currently no scientific evidence that plant diseases are spread via tools. Wound paints or other surface treatments should never be applied to cuts as they hinder healing and don't prevent infection.

CUT ABOVE A HEALTHY BUD

When you shorten the length of stem or branch, by making what's known as a "heading cut", you sever its growing tip. This removes the apical dominance (see pp.166–167) it exerted and diverts nutrients to the side buds below the cut, promoting bushier growth. Regrowth after pruning only ever comes from buds, so each cut needs to be made just above a healthy bud and as close to it as possible because all tissue above that bud will be compartmentalized as part of

BUD ARRANGEMENT varies between woody plant species and affects growth after pruning. Cut above an alternate bud to trigger growth in the desired direction. Both opposite buds grow equally after pruning.

OPPOSITE
Buds are arranged in pairs along the length of a stem.

ALTERNATE (LADDER)
Buds are staggered at intervals along two sides of a stem.

ALTERNATE (SPIRAL)
Single buds are arranged in a spiral pattern around a stem.

the plant's response to injury and left to die (see pp.166–167). This dead wood creates a fertile home for infection, so stumps above buds need to be kept as small as possible.

To direct new growth after pruning, cut just above a bud that faces in the desired direction. Gardeners usually prune to an "outward-facing" bud, pointing away from a plant's centre, to avoid branches becoming overcrowded and congested (see pp.164–165). This is simple where buds are staggered at intervals along either side of a stem in an "alternate" arrangement. However, some plants, such as *Clematis*, have pairs of buds directly opposite one another on the stem; here you can cut just above the pair to produce two new shoots, or cut at an angle to remove one bud and direct new growth.

It is widely recommended that heading cuts should be angled away from alternate buds at roughly 45°, so that water runs off and the stub above the bud is short, helping to prevent fungal infection. This is a myth, however, because an angled wound is larger than one made perpendicular to the stem, making it harder to repair. Research shows that a clean, flat cut just above a bud will heal faster and is the best insurance against infection.

CARE FOR THE BRANCH COLLAR

Thinning cuts are made to reduce the overall number of branches, removing dead or damaged limbs, reducing congestion, and allowing air and sunlight in to the centre of plant (see pp.164–165). This can also give trees and shrubs a tidier, more defined structure.

For many years, it was thought that cutting a branch flush with a tree's trunk would aid healing. We now know that the slight bulge at the base of a branch, called the "collar", is loaded with cells designed to combat injury by quickly churning out anti-fungal chemicals, and slowly expands to envelope and protect the pruning scar. Sawing off a branch at the trunk not only leaves a large gash, but also shreds the collar, sabotaging its ability to heal. Instead, cut vertically just outside the collar, or if it isn't visible, prune to leave a short stump.

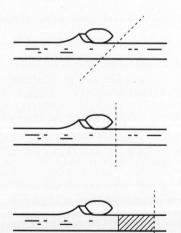

PRUNING CUTS should be made with care because their location and angle will affect the plant's subsequent ability to heal and to fend off infection.

ANGLED CUT
Pruning at 45°, angled away from the bud, is often recommended, but creates a larger wound.

STRAIGHT CUT
Cutting perpendicular to the stem, immediately above a bud, makes a small, fast-healing wound.

LONG STUB
Making a cut too far above a bud leaves a redundant section of stem that will die back.

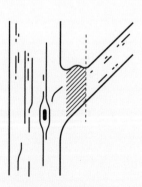

BRANCH COLLAR
Making a vertical cut just outside a branch's raised collar creates a smaller wound and aids healing.

SHOULD I TRAIN A FRUIT TREE AGAINST A WALL OR FENCE?

Most gardeners would love to have a fruit tree, but worry that they don't have the space. The solution is to train trees against walls, fences, or other vertical structures, where they will be both fruitful and beautiful.

The growth of young trees can be moulded (termed "training"), by bending and fixing their soft shoots into position before they stiffen. Training a tree against a flat surface or along wires and controlling its growth using pruning has a lot going for it: trees take up minimal space; branches are easier to reach for pruning and fruit picking; blossom and fruit can be protected from frost and birds; and the decorative forms make an attractive garden feature year-round. In the shelter of a south-facing spot, fruits develop and ripen faster thanks to the wall or fence reflecting heat, speeding up ripening chemical reactions.

WHY TRAINED TREES ARE FRUITFUL

Fruit trees produce two types of buds: those destined to become new shoots and leaves, and those that will form flowers. Buds on upward-pointing branches are awash with growth-suppressing auxin flowing down from the "apical bud" at the growing tip (see pp.164–167), forcing them to develop into slender leaf buds only, which are on standby to sprout into fresh shoots should the growing tip be damaged or lost. The buds on branches trained sideways have little or no auxin coming from above to silence them and so can produce more shoots able to develop fat flower buds. This, combined with the influence of other substances that induce flowering, means that branches trained away from vertical tend to be more fruitful, hence trees trained against walls or fences usually produce plenty of blossom and heavy crops.

PRUNING AND TRAINING

With patience you can train and carry out "formative" pruning to shape a young fruit tree yourself, or one can be bought partly or fully trained. Most fruit trees grown for training will have been grafted onto a semi-dwarfing or dwarfing rootstock (see pp.90–91) to keep their naturally vigorous growth in check and promote fruiting. Once a tree has its final form (see right) careful pruning is vital to control its growth and flowering. Normally you would prune free-standing trees in winter, when their starch reserves are high, to stimulate growth (see pp.168–169), but wall- and fence-trained trees are often best pruned in summer, when their sugar reserves are low, to limit new growth.

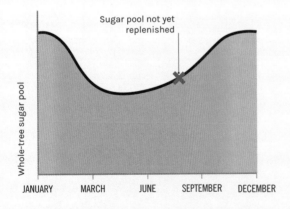

When to prune to restrict growth Summer pruning keeps trees compact because time for new growth before autumn is limited and the sugar stores to fuel growth are low.

KEY

Winter - dormant period

Growing season - repair and regrowth

— Level of sugar pool

✗ Best time to prune to restrict growth

Training fruit trees

Fruit trees can be trained in many forms. Select one to suit your space and the characteristics of each variety. Forms with branches 45–60 degrees from vertical produce even growth and are easiest to maintain.

FRUIT TREE FORMS

Training manipulates the flow of sideshoot-suppressing auxin from the main growing tips. Angling branches away from vertcial reduces auxin flow and shoot growth which, combined with other factors, allows more flower buds to form and yield fruit.

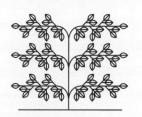

Vertical cordon
Ideal for small spaces
Suits: apples, pears, plums

Fan
Good for any location
Suits: apples, pears, figs,
plums, peaches, cherries

Espalier
Best for a high wall or fence
Suits: apples and pears

AUXIN FLOW FROM BRANCH TIP

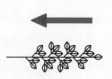

HIGH near growing tip
LOW further down trunk

MORE EVENLY SPREAD
along length and sides

LOW along branch
(but auxin from vertical shoots
pools along its base)

EFFECT ON SHOOT GROWTH AND FLOWER BUD FORMATION

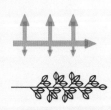

Top sideshoots vigour - **LOW**
Lower sideshoots vigour - **HIGH**
Flower bud formation - **LOW**

Sideshoot vigour - **MEDIUM**
Flower bud formation - **MEDIUM**

Top sideshoots vigour - **HIGH**
Lower sideshoots vigour - **LOW**
Flower bud formation - **HIGH**

WEAKENED

VARIEGATED LEAVES CONTAIN LESS CHLOROPHYLL AND MAKE LESS SUGAR TO POWER GROWTH

Pale-edged leaves make *Acer platanoides* 'Drummondii' (variegated Norway maple) a striking garden tree, but shoots frequently revert to produce plain green foliage.

ROGUE CELLS

A SINGLE MUTANT STEM CELL IN ONE LAYER OF THE MERISTEM GIVES RISE TO THE LEAF'S PALE EDGE

One mutant cell

Divided

Divided further

WHY IS MY TWO-TONE PLANT TURNING GREEN?

Two-tone (variegated) leaves are a beautiful contrast in a sea of plain green. Sometimes these prized plants sprout an odd shoot bearing plain green leaves. This is termed "reversion" and is a curious case of shape-shifting genetics.

Variegated plant varieties often originate from a stem with strange bicoloured leaves (called a "sport") that appeared on an otherwise green-leaved plant. A keen-eyed grower probably spotted this stem, snipped it off, and grew it on as a graft (see pp.90–91) or a cutting (see pp.184–185), eventually marketing it as a new variety.

AN ATTRACTIVE MUTATION

As far as a plant is concerned, however, variegated leaves are defective, caused by a partial genetic mistype in one of the rapidly dividing "stem cells" within the growing tip (meristem) (see pp.136–137). That one abnormal cell is pale in colour, since it is unable to produce chlorophyll, and as it further divides during growth, it produces a whole layer of cells within the leaf with the same error.

These chance mutations are sometimes "unstable", meaning they can occasionally be corrected during growth, causing green leaves

to unexpectedly appear on variegated plants. This is called "reversion". Because these genetically superior green leaves have a full complement of chlorophyll, they make more sugar than their two-tone neighbours and will grow more vigorously, eventually coming to dominate and turning the plant back to plain green. For this reason, any green-leaved reversion is best pruned out promptly. Occasionally, particularly on variegated hollies, a ghostly shoot with pure white or yellowish leaves may appear. Without any chlorophyll, these are even weaker than variegated leaves and never vigorous, so need not be snipped off.

FAST-GROWING SUCKERS

Green-leaved shoots can also appear on grafted plants (see pp.90–91). These "suckers" arise from the rootstock, below the graft, either from below soil or close to ground level. Suckers are effectively the rootstock portion of the plant trying to reassert itself and are often easy to recognise because they have slightly different leaves, and sometimes different flowers and fruit to your chosen variety, which grows from the upper "scion" portion of the plant above the graft. As with reversion, suckers are faster growing than the rest of the plant and, if left undealt with, will eventually take over.

Any shoot that grows from below the graft should be removed, by pruning it off close to the trunk. If it has come up from below ground level then it is best to pull it off, as this helps tear away any hidden buds at its base that may regrow, but if it's too tough then dig down and cut it off.

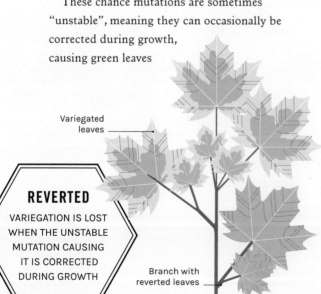

Variegated leaves

REVERTED

VARIEGATION IS LOST WHEN THE UNSTABLE MUTATION CAUSING IT IS CORRECTED DURING GROWTH

Branch with reverted leaves

CAN I COLLECT MY OWN SEEDS TO RAISE NEW PLANTS?

A plant's purpose is to reproduce and most garden plants will bear seeds if nature is allowed to take its course. It's easy to collect these seeds, but what grows might not be exactly what you bargained for.

Plants that self-pollinate (see p.88), including tomatoes and nasturtiums, are the easiest to save seeds from because they only need a single plant to reproduce. Many garden plants lack this "selfing" survival trick. Some avoid self-pollination using an inbuilt genetic check where only pollen from the same species that carries different DNA is accepted. Others structure flowers to prevent self-pollination or have male and female flower parts that mature at different times (see below), meaning another plant of the same species must be nearby to produce seed.

SEEDS NOT TO SAVE

It is not usually worthwhile collecting seed from plants labelled as F1 hybrids (see pp.88–89). They will not "come true" as the offspring (the F2 generation) won't have the same genetic purity as their parents and so have a range of characteristics, which may not make them good garden plants. Wind or insect pollination (open-pollination) is a complete lottery, which is why named varieties (cultivars) of perennials, trees, and shrubs are usually grown from cuttings and grafts (see p.90

KEY

- Plant
- ♀ Female flower parts
- ♂ Male flower parts

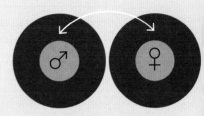

HERMAPHRODITE PLANTS	MONOECIOUS PLANTS	DIOECIOUS PLANTS

Route to pollination

A single plant can produce seeds if it can self-pollinate, but many need at least two plants of the same species to reproduce.

Single plant produces seeds
Flowers have both female and male parts and can self-pollinate. Common in garden plants.

TOMATOES (Solanum lycopersicum)

COLUMBINE (Aquilegia)

Single plant produces seeds
Flowers of different sexes form on the same plant. These can pollinate each other.

SQUASHES (Cucurbita)

SWEETCORN (Zea mays)

Multiple plants needed
Individual plants bear flowers of only one sex. Cross-pollination is essential Only female plants produce seed.

HOLLY (Ilex aquifolium)

ASPARAGUS (Asparagus officinalis)

and pp.184–185) to make sure they are genetically identical to the original plant.

COLLECT FOR SUCCESS

That said, open-pollinated seeds can give you excellent garden plants. Letting plants reproduce naturally and collecting their seeds can produce lots of plants, while offering the exciting possibility of a seedling appearing with fantastic looks or abilities. Propagate this with cuttings and you could create a new cultivar (see pp.180–181).

The seeds of many woody plants, annuals and biennials, are often worth saving, as are those from soft-stemmed (herbaceous) perennials that are difficult to grow from cuttings or don't easily divide (see pp.182–183). Seeds from open-pollinated varieties of vegetables can also be saved,

including "heirloom" or heritage varieties, so named because their seed has been saved for at least 50 years. Isolate vegetable varieties that are capable of cross-pollination, including beetroot or broad beans, by not growing them close to other related crops, to preserve their unique qualities.

WAYS TO SAVE

Collect seed as soon as it is "ripe". Seedheads, which include capsules, pods, and fruits, generally start out green and are ready when they are brown and dry, or when the fruit or berry has coloured. Cut off a whole seedhead and leave it to ripen and dry fully in a paper bag, before sorting the seed from pieces of capsule and storing somewhere cool and dry (see p.178). Remove seeds from ripe fruit, wash off any pulp, and dry before storing.

RIPE AND READY
WAIT UNTIL SEEDHEADS ARE BROWN AND DRY TO ENSURE SEED IS MATURE

Opium poppy *(Papaver somniferum)* seedheads each produce hundreds of tiny seeds for sowing.

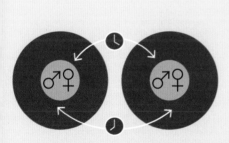

DICHOGAMOUS PLANTS

Multiple plants needed
Male and female parts within flowers mature at different times, so only cross-pollination is possible.

CUCKOO PINT *(Arum maculatum)*

ORNAMENTAL SAGES *(Salvia)*

WHAT'S THE BEST WAY TO STORE SEEDS?

Seeds contain a living cargo that is capable of erupting into new life months or years after falling. Most are easy to keep to sow in years to come, but some need some special care.

A living seed contains a tiny plant embryo that breathes oxygen and slowly ekes out its fat reserves. Most seeds contain enough food to last between two and five years, but must also be kept dry and at a steady, cool temperature. Low moisture prevents mould (fungal infection) while cool conditions slow the breakdown of fat reserves (see p.74). Before storing, make sure seeds are fully dry and mature (rather than soft and green), by leaving collected seed to dry indoors.

KEEPING SEEDS COOL AND DRY

Place clean, dry seeds in labelled paper envelopes and store in a sealed airtight plastic box or glass jar. Add a packet of silica gel, since this "hygroscopic" substance pulls any remaining moisture from the air. Powdered milk or oven-dried rice also do this job if 2–3 heaped tablespoons are sprinkled between layers of paper kitchen towel in the container. Seeds can be kept in the fridge at 3–5°C (37–41°F) if you intend to keep them for long periods, but also remain viable at a cool room temperature for sowing the following year. Seed banks freeze expertly dried seeds, but this is not recommended at home.

A few seeds do not store well and should be sown as soon as they fall from the plant – hellebores for example. Magnolias, oaks, and many tropical plants produce "recalcitrant" seeds, which are trickier to store because they will not germinate if they dry out. They remain viable for around three months kept in moist sand in a sealed polythene bag.

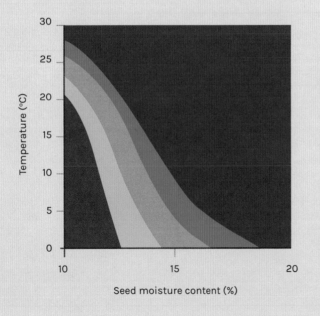

KEY

- Best long term storage
- Short term seed storage
- Fungal growth possible
- Fungal growth likely
- Rapid fungal growth and seed death

Cool, dry conditions are ideal
Thoroughly dried seeds will store well at up to 21°C (70°F). More moisture and warmth means seeds won't stay healthy for as long.

HOW LONG DO SEEDS REMAIN VIABLE?

Like a sleeping passenger frozen in time, the tiny living plant inside a seed can survive for months or millennia. How long depends on the plant species, and how the seed is collected and stored.

Some plants evolved to produce seeds that endure lengthy droughts and natural disasters, and although the seeds of most garden plants aren't viable (able to germinate into a healthy plant) for nearly as long, they often last for several years.

FACTORS AFFECTING VIABILITY

A seed's stamina hinges on the plant's genetics, the conditions in the year it formed, its moisture content, and storage temperature. High temperatures and high moisture content are the death knell for most seeds, which is why they are usually best kept in cool, dry conditions (see left). Stored well, most seeds remain viable for 2–5 years. Those best equipped to last have a strong protective seed coat, impregnated with powerful defensive chemicals, and effective molecular machinery to repair DNA damage accumulated over time. The small seeds of annual plants are often long-lived, for example, field poppies (*Papaver rhoeas*) can survive at least 8 years in the soil, as was so vividly shown in northern France at the end of the First World War.

TEST BEFORE SOWING

To find out whether a batch of seeds is still viable, test a small sample. Count out seeds and place them on damp paper towel in a polythene bag. Place somewhere warm and light and ensure the paper remains damp. If more than half germinate within two weeks then the seed is probably worth sowing.

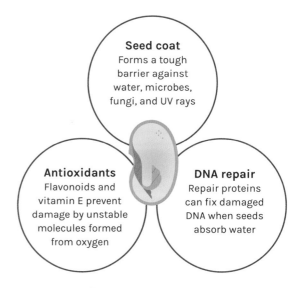

Self preservation Dormant seeds slow life processes to conserve energy, but are still able to prevent and even fix damage to help ensure successful germination.

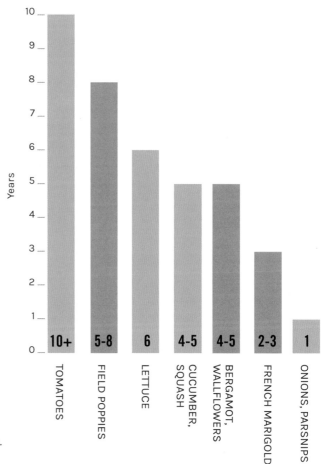

KEY

Seed longevity in years
The viability of some seeds drops much faster than others. Don't save or buy lots of seed that won't keep well.

COULD I BREED MY OWN PLANT VARIETY?

Plant breeders are forever wowing gardeners with new varieties, each boasting features they hope will prove irresistible. But with a little knowledge, lots of patience, and an eye for detail, any gardener can produce their own exciting new variety.

———————

When one plant is pollinated by another, each of the offspring could potentially become a new plant variety. Cross-pollinated offspring are a jumbled, genetically unique blend of mum and dad. Usually they aren't very different from their parents, but any that stand out, perhaps with distinctive flower colour, sweeter fruit, or resistance to a disease, are candidates to become a new variety. When a plant breeder finds or creates a plant with such a feature, they will multiply it by propagating it (see pp.136–137 and pp.176–177) and give it a unique name (see pp.64–65), making it a cultivar ("cultivated variety").

HOW ARE NEW CULTIVARS MADE?

Although some new cultivars are discovered as unusual "sports" on existing plants (see pp.174–175), most are the result of sexual reproduction through cross-pollination (see p.88). Plant breeders are expert at deliberate cross-pollination (hybridization) of plants, and use large glasshouses or fields to raise seedlings. You can do it on a smaller scale, and many successful cultivars have been created by amateur gardeners. There are two main ways of going about it, which can overlap.

THE SELECTION METHOD

Allow lots of plants to grow and reproduce, and you may find an exciting new cultivar among their seedlings. If you find a distinctive perennial, you could divide it or take cuttings to make many identical clones (see pp.182–185). But to collect seed and continue developing your cultivar, you

need to label your interesting specimen and keep it isolated (see right), so that it can't cross-pollinate with its "normal" neighbours. These flowers should then be self-pollinated, or in plants that may not self-pollinate, such as many from the daisy family (Asteraceae), allowed to cross-pollinate with a very similar plant (see p.176).

By growing these seedlings, weeding out those without the desired trait, and allowing those with it to cross-pollinate each other, you can create plants that produce the same features generation after generation (termed "breeding true").

DELIBERATE CROSS-POLLINATION

A more precise way is to deliberately cross-pollinate (hybridize) two plants with qualities that you want, such as red flowers and small size. To do this, first snip out the male pollen-topped stamens from a red-flowered plant (to prevent self-pollination) and then brush its female stigmas with pollen taken from a small plant. Do this vice-versa, using different flowers on the same plants, so that seed can be collected from each.

These seeds will produce the first (F1) generation (see p.89) of your hybridization. The seedlings may all be similar because the genetic code from one parent can sometimes be suppressed by that from the other, but if you allow these F1 plants to cross-pollinate, then traits from supressed (recessive) genes should reappear in the next (F2) generation. Sometimes cross-pollinating F1 plants with the original parent plants (termed back-crossing) can also reveal your desired trait.

Sport

Shoots with unusual characteristics, such as leaf variegation, can appear spontaneously. Vegetative propagation (see pp.136–137) will produce more plants with this attractive new trait.

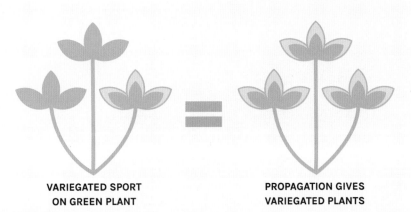

VARIEGATED SPORT ON GREEN PLANT

PROPAGATION GIVES VARIEGATED PLANTS

Selection

If you grow lots of plants, sexual reproduction may chance on something special. Isolate this plant, collect its self-pollinated seed (see p.88), and raise seedlings that should exhibit this feature.

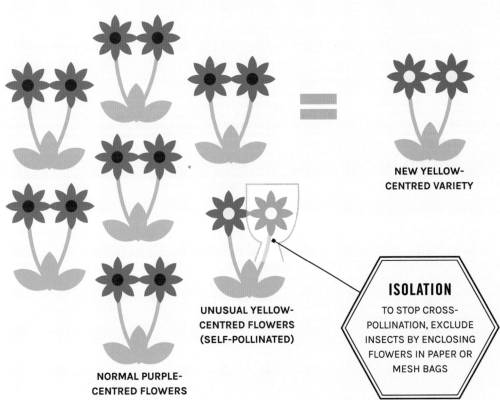

NEW YELLOW-CENTRED VARIETY

UNUSUAL YELLOW-CENTRED FLOWERS (SELF-POLLINATED)

NORMAL PURPLE-CENTRED FLOWERS

ISOLATION

TO STOP CROSS-POLLINATION, EXCLUDE INSECTS BY ENCLOSING FLOWERS IN PAPER OR MESH BAGS

Cross-pollination

Select two plants with features that you want to combine and cross-pollinate their flowers. Here, tomatoes with different fruit colour and shape are crossed to form a new cultivar.

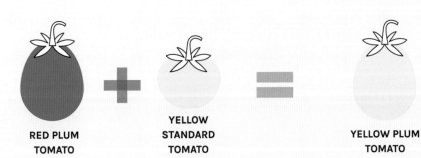

RED PLUM TOMATO

YELLOW STANDARD TOMATO

YELLOW PLUM TOMATO

IF I SPLIT UP A PLANT, WILL IT DIE?

Picture the scene: you thought you were digging out a stubborn weed, but in fact you've sliced through a clump of your favourite perennial that was sending out new shoots. Despair not! Marvel instead at the incredible regenerative powers of plants...

For very many plants, being severed down the middle need not be a death sentence. Far from it, in fact: thanks to plants' innate abilities to clone themselves like Dolly the Sheep (see pp.136–137), herbaceous (non-woody) plants that grow in clumps can be dug up and split into several sections, which will continue to grow as brand new plants.

OUT OF ONE, MANY

Plants such as *Rudbeckia* (black-eyed Susan) grow outwards by multiplying themselves sideways, each stem a clone of the next, with a ball of embryonic (meristem) tissue at its base (see pp.136–137) and a root beneath. Each stem-root unit can therefore be teased apart and grown as an independent plant. Not only will dividing a plant give you extra free plants, but each will be given a new lease of life, since each section can now wallow in a fresh supply of water, light, and nutrients without competition. And you needn't worry about leaving the plants with injuries: immediately after a cut is made, most plants send out a chemical "wounding signal" that reverberates throughout the

SHOOTS

NEW LEAVES AND STEMS ARISE FROM BUDS IN THE PLANT'S "CROWN", AT OR JUST ABOVE SOIL LEVEL

Shoots grow from embryonic cells in the apical and axillary meristems

Developing leaves protect shoot tip

Heucheras are always more attractive as young plants and it's easy to cut spritely sections with strong roots and shoots from established specimens to create a new generation.

plant, triggering production of defensive substances, such as antibiotics, to ward off infection. As with pruning, clean straight cuts will speed callus formation over the wounded surface (see pp.166–167).

LIGHT LIFTING

All clump-forming perennial plants, such as hostas (*Hosta*), primroses (*Primula*), asters (*Aster*), and arum lilies (*Zantedeshcia*), can be divided by easing them up from the ground – or "lifting" in gardening parlance – using a garden fork to minimize soil disturbance, and being careful to damage roots as little as possible. Healthy roots lead to a healthy division – and it is the microscopic "root hairs" (see pp.104–105) that are critical to the plant's survival when it is replanted. The more that soil is disturbed around the roots, the more of these tiny, delicate root hairs that will die. Plants with lots of stringy, fibrous roots, such as primroses, can either be teased apart by hand into smaller clumps or, if large and bulky, as with hostas, sliced into two or more pieces, provided each piece has at least one complete stem or bud.

Older plants may have dead patches in the centre or dying stems that have been starved of nutrients and water by the younger, more vigorous outer stems. When plants are lifted, dying or diseased areas can be cut out and added to the compost pile to recycle their nutrients (see pp.190–191).

Plant sections can be simply replanted into the same hole, possibly with added compost, and can also be planted in new locations. Make sure the planting holes are big and deep enough for the roots to spread out comfortably. Coiling the roots into a tight hole will cost the plant energy as it will need to grow extra roots to reach out into fresh soil. After backfilling around roots with soil, press down gently around the plant ("firm in") and water generously so that there are no large, air-filled gaps around the roots, which ensures the root hairs will regrow.

ROOTS

IT'S VITAL THAT EACH DIVISION HAS SOME WELL DEVELOPED ROOTS TO SUPPLY LEAVES WITH NUTRIENTS AND WATER

Split times

TO KEEP PLANTS VIGOROUS, IT IS GENERALLY RECOMMENDED TO LIFT AND DIVIDE HERBACEOUS PERENNIALS EVERY THREE TO FIVE YEARS, when they start to become overcrowded, or sooner if there is a bare patch in the centre. Dividing can be done at any time but is usually best done outside of a plant's main growing season, so that the underground plant parts are laden with reserves, but shortly before plants put on new growth. For most, this will be in spring when the soil starts to warm, although some prefer an autumn division.

CAN I GROW ANYTHING FROM CUTTINGS?

Plants have an extraordinary ability to grow roots and regenerate from almost any body part. Gardeners take advantage of this by using cuttings to make identical copies of favourite plants, but not all can be propagated in this way.

Most cuttings are taken simply by snipping a plant stem. This causes a hormone called jasmonate to set off a healing response, readying the area for root growth. As auxin, the powerful plant hormone, travels down to the wound from the shoot tip, stem cells (see p.137) at the cut multiply to form "adventitious roots" (see p.105). If kept damp, these roots will grow and help form a new plant. Not every plant can be grown from cuttings, however. Ferns, grasses, orchids, and other plants classed as "monocots", lack the internal vein structure with surrounding stem cells needed to sprout new roots.

WHAT TO PROPAGATE WHEN

Stem cells in young, soft stems form adventitious roots most readily and cuttings from a few soft-stemmed plants, such as *Pelargonium*, even root easily in a glass of water. Many other plants will grow from a stem cutting pushed into soil or potting compost (see p.193), and for perennials and deciduous shrubs "softwood" cuttings from young stems in early to midsummer often work best. Climbers and evergreens root better from "semi-ripe" cuttings, with a soft tip and a woodier base, taken from late summer to early autumn. "Hardwood" cuttings are taken from woody stems of deciduous shrubs from mid-autumn to early winter. They take much longer to root, but lose little water while dormant and leafless.

Incredibly, a few plants can even form plants from leaf cuttings or small sections of root, if they are inserted into damp potting mix.

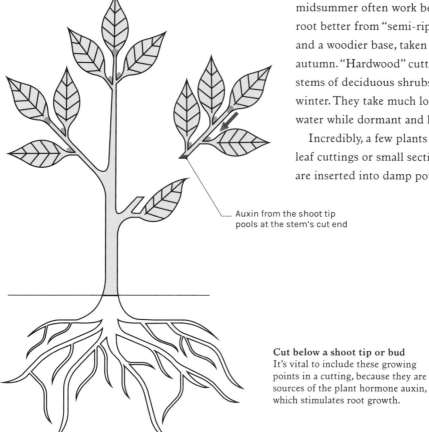

Auxin from the shoot tip
pools at the stem's cut end

Cut below a shoot tip or bud
It's vital to include these growing points in a cutting, because they are sources of the plant hormone auxin, which stimulates root growth.

KEY

▶ Auxin flow

■ Auxin accumulates

● Buds

IS THERE A TRICK TO SUCCESS WITH CUTTINGS?

As anyone who has tried will know, growing plants from cuttings can seem like a lottery. But careful preparation, thorough hygiene, and a sprinkling of science increase the odds of success.

———————

Healthy cuttings come from healthy plants, and so make sure the parent plant is well-fed, hydrated, and free from disease.

MAKE A HEALTHY START

Moisture is essential, since severed stems will quickly dry out. Place cuttings in a polythene bag and always plant promptly. Cuttings should usually be 10–15cm (4–6in) long – longer and they will dehydrate, shorter and they won't contain enough resources to form roots. Hormone rooting products contain synthetic auxin, which accelerates root formation when dabbed on the cut surface and increases success for cuttings that are trickier to root.

GOOD SURROUNDINGS FOR ROOTING

Use clean pots, sterilized tools, and fresh compost to reduce chance of disease. Use a compost that drains well or mix your own using 50 per cent perlite or vermiculate. Push cuttings into compost, water well, then cover the pot with a clear plastic bag supported on sticks to stop condensation touching leaves. A heated propagator supplies high humidity and warmth at the base, both of which promote rooting. Leave cuttings of hairy or silver-leaved plants uncovered or else they may rot. Keep in a bright place, out of direct sun, to prevent overheating. New leaf growth shows that roots have formed.

Care at every stage

Keep softwood cuttings moist by taking them in the morning or in cool conditions and putting them in a polythene bag. Prepare and plant as soon as possible.

KEY

Water evaporating from leaves and compost is trapped

▶ Auxin flow

■ Auxin accumulates

▷ Water

● Buds

CUT
Use a sharp knife to cut just below a leaf joint, where auxin will collect for better rooting.

PREPARE
Remove leaves to reduce water loss, but leave enough to provide sugar for growth.

PLANT
Push into compost, water, and cover with a clear bag to raise humidity and stop cuttings drying out.

WHAT IS LAYERING?

Stems will often form roots if they are pulled downwards and kept in contact with soil. Layering is a simple method of plant propagation that takes advantage of this ability, creating a new plant that can be severed from its parent and replanted.

Layering is successful for many woody plants, including hazel (*Corylus avellana*), *Wisteria*, and some *Magnolia*. It's best carried out in spring or autumn, when soil is warm and moist.

TRY SIMPLE LAYERING

First, bring a pliable, young stem down into contact with soil. Bury a short section with the leafy tip pointing upwards, secure it with a bent piece of thick wire, and the dark, damp surroundings will trigger the meristem tissue on the underside of the stem (see pp.136–137) to produce roots. Making a slanted cut about halfway through the stem at the point where it is buried will speed up root growth, since it channels nutrients and the growth hormone, auxin, to the injury. Once roots are established, the

connection to the parent plant can be cut and the new plant replanted.

ALTERNATIVE METHODS

Some plants, including blackberries, naturally layer by rooting easily whenever the tips of their arching stems touch soil. Tip layering is an easy way to generate more plants, but it happens so readily that stems need to be tied to supports to control their spread. Others, such as bugle (*Ajuga reptans*) and strawberries (*Fragaria*), grow special horizontal stems called runners (see pp.182–183) with plantlets along their length, which rapidly root into soil.

"Air layering" is a more complex method where a stem is nicked so that roots form when it is wrapped in plastic filled with moist compost or moss.

Two layering techniques
Simple and tip layering are easy ways to create new plants from those already established in your garden.

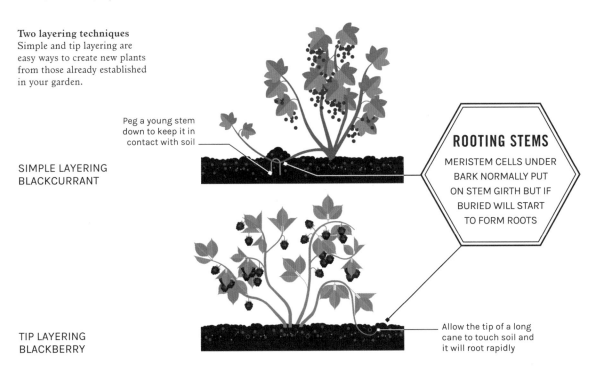

Peg a young stem down to keep it in contact with soil

SIMPLE LAYERING
BLACKCURRANT

ROOTING STEMS

MERISTEM CELLS UNDER BARK NORMALLY PUT ON STEM GIRTH BUT IF BURIED WILL START TO FORM ROOTS

TIP LAYERING
BLACKBERRY

Allow the tip of a long cane to touch soil and it will root rapidly

HOW CAN I MAKE NEW PLANTS FROM BULBS, CORMS, RHIZOMES, AND TUBERS?

While most plants reproduce by flowering and shedding seeds, some multiply underground bulbs, corms, rhizomes, and tubers (see pp.96-97) to effectively clone themselves. This makes it easy for gardeners to produce new plants.

Bulbs and corms reproduce by growing identical baby copies (offsets), sometimes termed bulblets and cormlets, from the base of the mother bulb. These offsets grow into new plants, forming a clump where originally a single bulb was planted.

EASY MULTIPLICATION BY DIVISION

By digging up the underground part of the plant ("lifting"), offsets can simply be separated from the larger parent bulb and replanted. This is best done when the leaves have yellowed and wilted, indicating that leaf sugars and nutrients have been pulled into the underground storage organs in preparation for dormancy (see pp.152–153). Most bulbs are best lifted and divided every three to five years, when the parent bulb is not flowering as vigorously.

MORE COMPLEX TECHNIQUES

Astonishingly, some storage organs can be sliced or broken into pieces and regrown as separate plants. Rhizomes and tubers can be cut into sections containing at least one bud, and then replanted.

The loosely packed, fleshy "scales" of scaly bulbs, such as lilies (*Lilium*), can be broken off and kept in a sealed bag of moist compost at 20°C (68°F) until bulblets sprout from the base, at which point the bulblets can be planted. Some bulbs, such as snowdrops (*Galanthus*), can be propagated by slicing them vertically into orange-like segments (called chipping) then treating them in the same way as scales.

Plant one garlic clove (which is a bulblet) and it will multiply into a whole bulb containing eight or more cloves in one growing season (8–9 months).

WHAT IS COMPOST AND HOW DOES IT FORM?

Compost is a brown, crumbly material, with an earthy smell, produced when anything that was once living (organic matter) has decomposed. It's a wonder-food for soil that supercharges plant growth, yet is easy to make from ordinary garden and kitchen waste.

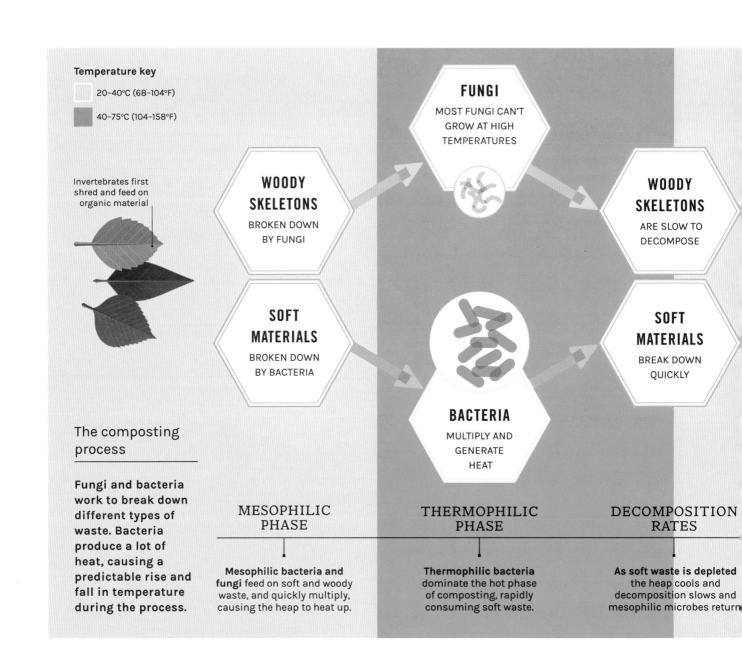

Temperature key

■ 20–40°C (68–104°F)

■ 40–75°C (104–158°F)

Invertebrates first shred and feed on organic material

WOODY SKELETONS
BROKEN DOWN BY FUNGI

SOFT MATERIALS
BROKEN DOWN BY BACTERIA

FUNGI
MOST FUNGI CAN'T GROW AT HIGH TEMPERATURES

WOODY SKELETONS
ARE SLOW TO DECOMPOSE

SOFT MATERIALS
BREAK DOWN QUICKLY

BACTERIA
MULTIPLY AND GENERATE HEAT

The composting process

Fungi and bacteria work to break down different types of waste. Bacteria produce a lot of heat, causing a predictable rise and fall in temperature during the process.

MESOPHILIC PHASE

Mesophilic bacteria and fungi feed on soft and woody waste, and quickly multiply, causing the heap to heat up.

THERMOPHILIC PHASE

Thermophilic bacteria dominate the hot phase of composting, rapidly consuming soft waste.

DECOMPOSITION RATES

As soft waste is depleted the heap cools and decomposition slows and mesophilic microbes return

Before you get stuck into making compost (see pp.190–191), it pays to understand the different jobs that fungi and bacteria have and how these change through the composting process. That way, you can feed and manage your heap to produce crumbly compost as quickly as possible and can recognise when it's ready to spread on your garden.

COMPOSTING ALLIES

Once insects, worms, slugs, snails, beetles, woodlice, and other hungry invertebrates have shredded and fed on large

pieces of organic material, bacteria and fungi go to work digesting whatever is left (what we usually call "rotting"). These two microscopic life forms are very different: fungi grow in long threads and release powerful plant-digesting chemicals that can break down woody materials; bacteria are much smaller, single-celled organisms that swarm and cluster, and typically feed on soft matter. Both are fuelled by carbohydrates – the sugars and fibrous components in plants. And they need nitrogen to make proteins and DNA, and so reproduce.

Only fungi can eat the tough, carbon-based "skeleton" that plants invest so much energy in building. It can take over two years for the tough lignin that gives wood and bark their strength to fully break down.

Bacteria, on the other hand, feast on soft, juicy materials, such as grass clippings or food scraps. They proliferate at astonishing rates, doubling in number every 30 minutes as they digest simple carbohydrates, ingest sugars, and hose up nitrogen from proteins, chlorophyll, and other plant pigments. These bacteria generate heat, which accelerates their growth and the decomposition process further.

DECOMPOSITION GENERATES HEAT

Given ample food and air, the activity of these "mesophilic" bacteria will cause temperatures in a compost heap to soar to over 40°C (104°F) in 24–72 hours, at which point other heat-loving ("thermophilic") bacteria and heat-resistant fungi go to work on the hardiest proteins, structural carbohydrates, and fats. As these fuels are used up the pile shrinks, bacterial growth slows, and the organic matter cools. It is important to "turn" a heap so air reaches bacteria and undecomposed material from the edges is moved into the centre to be broken down. Mesophilic bacteria once again take over to finish the "maturing" phase of the process, along with invertebrates, fungi, and unusual fungi-like bacteria, called actinomycetes, which emit the earthy aroma that indicates a job well done. Allow time for this final phase before spreading your compost.

Compost is nourishment for soil, supporting the organisms of the soil food web (see pp.44–45) that feed plant growth with a full range of plant nutrients which stay in the soil, rather than washing away, like those in many fertilizers. It also teems with more bacteria and fungi than healthy soils. Spread as a mulch, it boosts numbers of soil organisms, improves soil structure (see pp.42–43), and suppresses weed growth (see pp.46–47).

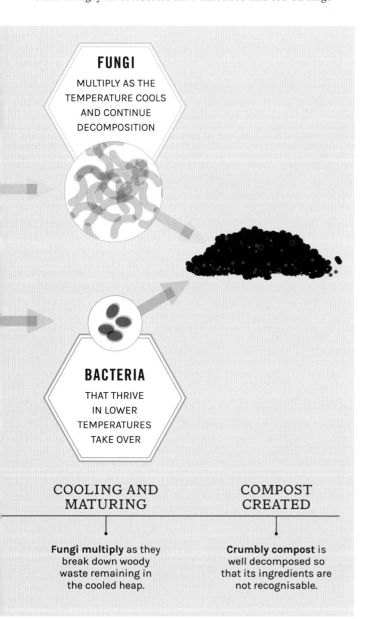

FUNGI
MULTIPLY AS THE TEMPERATURE COOLS AND CONTINUE DECOMPOSITION

BACTERIA
THAT THRIVE IN LOWER TEMPERATURES TAKE OVER

COOLING AND MATURING

COMPOST CREATED

Fungi multiply as they break down woody waste remaining in the cooled heap.

Crumbly compost is well decomposed so that its ingredients are not recognisable.

WHAT'S THE BEST RECIPE FOR HOMEMADE COMPOST?

There are as many ways to make compost as there are to cook an egg and, as with any recipe in the kitchen, getting the type and proportion of ingredients right will set you on the road to success.

A simple heap of garden waste will eventually be digested by countless creatures to produce rich compost. A more controlled method is to gradually add organic material to open bays (see right) or to a lidded "bin", which you can choose to suit your space and amount of waste. Larger heaps generate more heat (see pp.188–189) and will decompose faster, and adding a lid or cover keeps out excess water and pests. Rotating "tumbler" bins make light work of mixing a heap to add air. Place your heap on soil, or add a few spadefuls of soil with the first organic material, to inject a goodly dose of bacteria and fungi to kickstart the composting process.

FOOD FOR MICROBES

The speed at which finished compost is produced depends on the balance of foods present for fast-dividing bacteria and more sedate fungi. Autumn leaves contain lots of carbon to nourish fungi, but by themselves can take two or three years to fully decompose into "leafmould". Move this slow, fungus-driven composting into the fast lane by adding nitrogen-rich food for bacteria, such as grass clippings, fresh leaves, and weeds. But be careful, because an excess of high-nitrogen waste can lead to runaway bacterial growth, causing the temperature to soar to over 80°C (176°F), killing beneficial microbes, and even risking fire. The correct balance of carbon and nitrogen is key and a helpful way of thinking about it is to categorize nitrogen-rich waste as "green" and carbon-rich waste as "brown".

AERATION

AIR IS VITAL FOR GOOD COMPOST, BECAUSE THE DECOMPOSERS THAT FORM SWEET, CRUMBLY COMPOST NEED TO BREATHE

The ideal nutritional balance is provided by a mix of waste with thirty times more carbon than nitrogen (a ratio of 30:1). Because plant material is built around a carbon-based skeleton (see pp.188–189), this works out as a roughly even mix of green and brown waste by volume. This rule of thumb is calculated for garden waste only – other compostable material doesn't always fit neatly into the colour categories. Paper and cardboard are "brown", as you'd expect. Food scraps, coffee grounds, and animal manure, however, make great, nitrogen-rich foods for bacteria even though they aren't coloured green. Think of compostable waste as either food for fungi (often dry and lifeless) or for bacteria (usually fresh and containing moisture).

TURNING AND MOISTURE

Combining plenty of dry, bulky brown waste with dense, damp green waste allows air to flow and microbes to breathe. Mixing or "turning" a compost pile with a fork or in a compost tumbler, also adds air to reinvigorate and speed up decomposition. Do this at least once during the process, after the pile has heated up. The more a heap is turned (as often as weekly) the faster compost will be ready.

Moisture is also vital for microbes, and is usually plentiful when a heap is 50 per cent green waste, but may need to be added in dry climates. A wet heap lacking air becomes home to "anaerobic" (without-air) bacteria, which decompose waste slowly and release acids, alcohols, methane, and putrid gases.

Composting with two bays holds plenty of garden waste and makes turning easier as you move compost from a full to an empty bay.

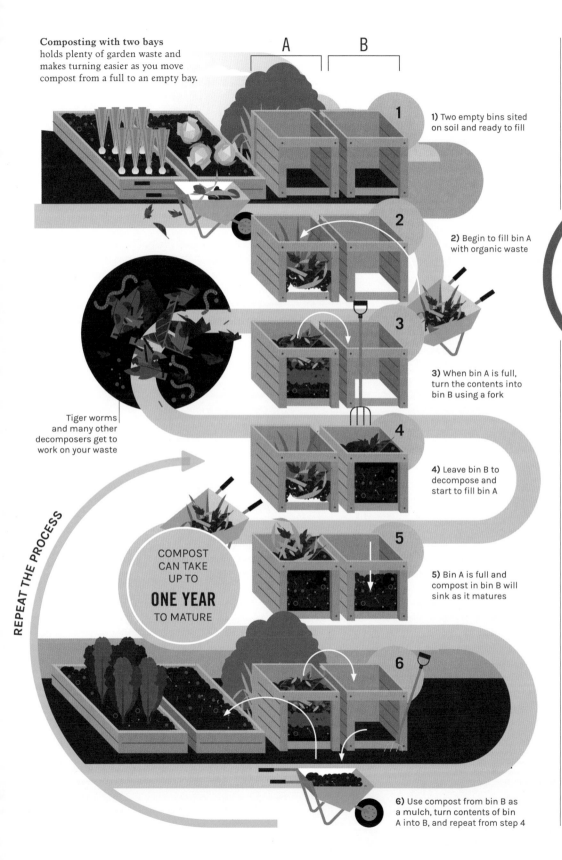

A   B

1) Two empty bins sited on soil and ready to fill

2) Begin to fill bin A with organic waste

3) When bin A is full, turn the contents into bin B using a fork

Tiger worms and many other decomposers get to work on your waste

4) Leave bin B to decompose and start to fill bin A

COMPOST CAN TAKE UP TO **ONE YEAR** TO MATURE

5) Bin A is full and compost in bin B will sink as it matures

REPEAT THE PROCESS

6) Use compost from bin B as a mulch, turn contents of bin A into B, and repeat from step 4

CARBON TO NITROGEN RATIOS

ALL ORGANIC WASTE CONTAINS CARBON AND NITROGEN IN VARYING PROPORTIONS. THE IDEAL BALANCE TO GIVE FAST, AEROBIC COMPOSTING IS:

30:1
CARBON   NITROGEN
(C)   (N)

MIX EQUAL VOLUMES OF GREEN AND BROWN WASTES TO ACHIEVE CLOSE TO THIS C:N RATIO AND MAKE RICH, CRUMBLY COMPOST.

"BROWN" WASTE

HIGHER IN CARBON (C:N)

DEAD LEAVES **60:1**

SHREDDED PAPER **100:1**

CARDBOARD **125:1**

STRAW **50:1**

CHIPPED WOOD **80–145:1**

SAWDUST **150:1**

"GREEN" WASTE

HIGHER IN NITROGEN (N:C)

GRASS CLIPPINGS **15:1**

COFFEE GROUNDS **20:1**

GREEN LEAVES **17:1**

VEG PEELINGS **20:1**

HAIR AND FUR **10:1**

MANURE **5–25:1**

SHOULD I AVOID PUTTING WEEDS AND DISEASED LEAVES IN COMPOST?

No-one wants to mulch their garden with homemade compost only to find a blanket of weeds emerging or disease taking hold, so just how careful do you need to be about adding weeds and diseased material to your heap?

Weeds are, by their very nature, highly efficient growers and rapidly accumulate nitrogen from soil in their leaves. They can often be appreciated in gardens for their wildlife benefits (see pp.46–47), but when they need to be removed they make a very valuable addition to compost.

Watch out, however, because weeds are resilient plants, capable of surviving composting, and regrowing if given the chance. Bindweed, dandelions, couch grass, and other perennials can regenerate from fragments of their thick, pale roots, which can survive in cool heaps. Leave these weeds to dry thoroughly in the sun before composting, so they have no hope of resurrection.

HOT COMPOSTING REDUCES PROBLEMS

The hard seed coats of weed seeds are impregnated with microbe-repelling poisons (antibiotics), which help them withstand composting bacteria and fungi. Where possible, avoid adding them to compost by pulling weeds up before flowers and seedheads develop. These seeds have little defence against the high temperatures in a hot compost heap (see pp.188–189), however: at 40°C (104°F), the proteins controlling their life processes (enzymes) start to break down and the higher the temperature the faster they will be killed. Many of the fungi, bacteria, and viruses that cause plant diseases (see pp.194–195) are less vulnerable to being cooked. Research shows that while most are eradicated at 60°C (140°F), some survive beyond 80°C (176°F) by cocooning themselves in egg-like spores which are impervious to heat. Diseased plants are therefore best not composted. Use a thermometer probe to check the temperature of your heap and the likelihood of weeds and diseases surviving.

40°C (104°F) ENZYMES IN WEED SEEDS START TO BREAK DOWN, KILLING THEM

60°C (140°F) MOST BACTERIA, FUNGI, AND VIRUSES ARE KILLED

80°C+ (176°F+) SOME DISEASE-CAUSING MICROBES SURVIVE PROTECTED IN HEAT-REISTANT SPORES

Compost heat tolerance
Large heaps and insulated hot bins generate and retain heat best, but it is prudent to avoid composting diseased material as some plant diseases will survive.

WHY DO I NEED TO LET MANURE ROT DOWN?

Leaving fresh manure to decompose ("rot down") makes it safer for plants and humans, because this process reduces levels of harmful bacteria and chemicals. It also helps to kill weed seeds and makes manure much easier to handle.

———

The digestive juices and bacteria-loaded guts of plant-eating animals break down plant material in a matter of hours. Since no digestive system extracts all of food's goodness, animal manure still contains half of the nitrogen, most of the phosphates, and nearly 40 per cent of the potassium (see pp.122–123) in the original food. This should make it an ideal plant fertilizer, but used fresh, it can cause problems.

ISSUES WITH FRESH MANURE

Weed seeds consumed are weakened but not killed by digestive acids and enzymes, and manure can be rich in harmful bacteria: not good for edible crops. Fresh manure has high concentrations of ammonia and urea, chemicals that can "scorch" plants, causing yellowing leaves as water is pulled from roots into soil. Manure also includes woodchip and straw, which draw nitrogen from soil as they decompose.

THE BENEFITS OF COMPOSTING

Left to stand for just a few weeks, enough urea and ammonia will evaporate from manure to make scorching unlikely. In three to six months, a pile of manure will decompose (helped by bacteria from the animal's gut), killing weeds seeds, breaking down bedding materials, and eradicating harmful bacteria in the heat of the process. "Well-rotted" manure crumbles like compost and loses its unpleasant smell.

POISONS CAN PERSIST

Source manure carefully, as it may be contaminated with a powerful weedkiller (aminopyralid) that has been sprayed on the fields that grew the straw or hay fed to animals or used as bedding. This herbicide kills broad-leaved plants (not grasses) and causes the growth of plants mulched with contaminated manure to be stunted and malformed.

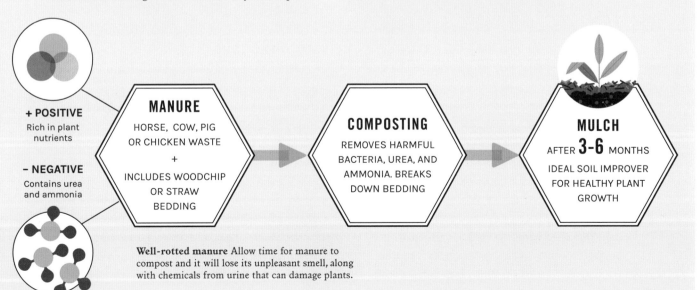

+ POSITIVE
Rich in plant nutrients

– NEGATIVE
Contains urea and ammonia

MANURE
HORSE, COW, PIG OR CHICKEN WASTE
+
INCLUDES WOODCHIP OR STRAW BEDDING

COMPOSTING
REMOVES HARMFUL BACTERIA, UREA, AND AMMONIA. BREAKS DOWN BEDDING

MULCH
AFTER **3-6** MONTHS
IDEAL SOIL IMPROVER FOR HEALTHY PLANT GROWTH

Well-rotted manure Allow time for manure to compost and it will lose its unpleasant smell, along with chemicals from urine that can damage plants.

WHAT SORT OF DISEASES DO PLANTS GET?

It's not just humans who can catch a nasty bug – plants can too. Understanding their ailments and the microscopic attackers that cause them can help us diagnose disease and nurse them back to health.

———————

Diagnosing an illness is not easy when your patient can't tell you where it hurts. Worse, the common symptoms of plant disease rarely point to one cause. For example, yellowing ("chlorosis") happens when leaves stop producing chlorophyll, and can be caused by a nutrient deficiency, unsuitable soil pH, or root damage, as well as infection. Yellowing between leaf veins, called "interveinal chlorosis", indicates iron deficiency if in young leaves, and manganese or magnesium deficiency if in older ones. Low potassium levels cause yellowing around leaf edges, while nitrogen deficiency shows as yellowing in older leaves. When such common causes have been ruled out, turn your attention to infections.

FUNGAL PROBLEMS

Fungi generally struggle to survive at human body temperature so rarely bother us, but they cause 85 per cent of plant infections. When certain fungi, called powdery mildews, amass on leaves, they form a visible, frost-like coating. This infection of the leaves's "skin" may cause a plant to lose leaves, but it will rarely succumb. Many fungi, such as grey mould, enter plants through a cut, wound, or insect bite. Nastier fungi can puncture or chemically fizzle through a plant's tough skin (called the cuticle). Black, brown, or yellow blotches on leaves and stems are the hallmarks of infection, where multiplying microbes are digesting and destroying tissue. "Rust" fungi invade leaves and stems, organising their thread-like growth (hyphae) into small orange-brown lumps, termed pustules. These are the fungus's "fruiting bodies", each bulging with

thousands of minute "seeds" (called spores), ready to erupt and spread to infect its next victim.

WHEN TO BLAME BACTERIA

Bacteria cover practically every surface on the planet, but just a handful cause plant disease. They often cause "leaf spotting" – dark spots or holes – where bacteria have multiplied and destroyed the leaf, surrounded by a yellow halo of injured tissue. They can also produce bulging bacteria-filled pustules, slow-growing, blackened patches on trees (called canker), as well as ugly blotches on fruit, and lumps on stems and roots, known as galls (see pp.198–199). Some bacteria, such as fireblight, enter plants through flowers or even leaf pores on a wet day, but most can only breach a plant's defences through an open wound or insect damage. Affected stems, bulbs, and roots can become slimy and disintegrate, giving off an unpleasant odour.

VIRAL INFECTIONS

Completing the list of undesirables are viruses: tiny particles of genetic code that infiltrate plants – often through an insect bite – to highjack cells and multiply. Symptoms are often vague, and include stunted, distorted, or weak growth, but also yellow, brown, or black streaks, flecks, or mosaic patterns on leaves. Viruses are named after the first plant they were discovered in, regardless of the plants they can actually infect. Cucumber mosaic virus, for example, can infect many plants, including spinach and celery. They rarely kill a host, but will weaken it over several years, while potentially spreading to nearby plants.

Causes and symptoms of plant diseases

The germs that plants pick up are not the same as those that make humans ill. We cannot catch tomato blight, any more than a gerbera can catch the flu. Nevertheless, the microscopic attackers are similar: fungi, bacteria, and viruses.

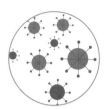

BACTERIA

FUNGI

VIRUSES

Form dark leaf spots with a yellow halo, bulging growths, and foul-smelling rots.

Produce powdery leaf coatings and pustules, or fast-spreading rots and moulds.

Distinctive streaked or patterned markings on leaves, and weak or distorted growth.

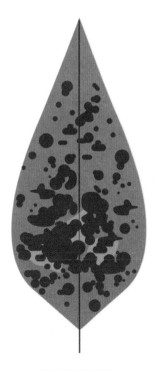

INFECT VIA:

- WOUND OR PEST DAMAGE
- FLOWERS
- LEAF PORES

INFECT VIA:

- WOUND OR PEST DAMAGE
- LEAF PORES
- PUNCTURING CUTICLE

INFECT VIA:

- INSECT PESTS
- CUTTING TOOLS
- HANDS

HOW CAN I STOP PLANTS BECOMING DISEASED?

Keeping infections at bay might seem impossible, but there are several simple and effective steps you can take: keep plants healthy with good growing conditions, avoid plants prone to diseases in your area, and plant resistant varieties where available.

Prevention is better than cure. Sickly plants pick up more infections, which is why good plant health is the cornerstone of preventing disease. Protecting plants from stressful conditions is the vital first step in an integrated approach to disease management (see pp.52–53).

USE DISEASE RESISTANT PLANTS

Some plants have natural resistance to disease, and others have been bred to fight off common pathogens (microbes or viruses that cause disease). Grown outdoors, most varieties of tomato succumb to late blight (*Phytophthora infestans*) during a warm, wet summer, but a cultivar with blight resistance genes, such as 'Mountain Magic', will be immune. Plants with disease resistance have a range of special abilities: they may produce anti-toxins to neutralize poisons released by fungi; have cells that lack the molecular "handles" that viruses latch on to; or might manufacture bacteria-killing antibiotics.

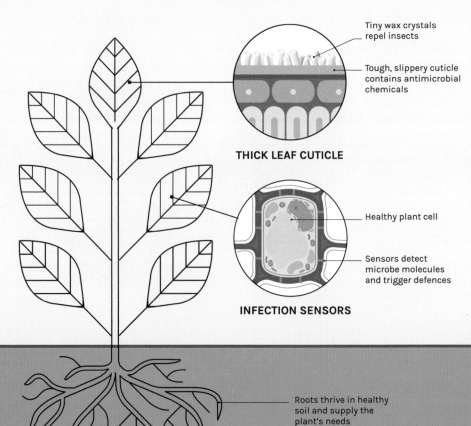

Tiny wax crystals repel insects

Tough, slippery cuticle contains antimicrobial chemicals

THICK LEAF CUTICLE

HEALTHY PLANT
Given ideal growing conditions, healthy soil, and some natural disease resistance, plants are able to sense and fend off most attacks. Thick leaf cuticles act as armour, while chemical defences fight off intruders.

Healthy plant cell

Sensors detect microbe molecules and trigger defences

INFECTION SENSORS

Roots thrive in healthy soil and supply the plant's needs

HEALTHY PLANTS FIGHT INFECTION

Plants can usually defend themselves if well cared for and planted in a suitable spot (see pp.58–59). Get off to a good start, and avoid bringing pathogens into the garden, by buying healthy plants, with no wilting, yellowing, or spots on their foliage. Mulch regularly to keep soil healthy, so that it can provide plants with the water and nutrients needed for strong immune systems (see pp.42–43).

Well-tended plants typically have a tough waxy skin (cuticle) over their leaves, which is slippery and contains antibacterial and antifungal chemicals. Vibrant green leaves also provide energy to fight intruders. In healthy plants, cell sensors detect distinctive molecules on bacteria and fungi and manufacture toxins tailored to that invader; leaf pores (stomata) snap shut to exclude microbes; the waxy cuticle and cell walls thicken; and infected cells produce virus neutralizing substances.

KNOW YOUR ENEMY

Find out which diseases affect which plants, and avoid those that are prone to infection. For example, many busy lizzies (*Impatiens*) suffer from downy mildew, hollyhocks are troubled by rust, and box can be ravaged by box blight. Ask local gardeners about the diseases they encounter and learn to recognise symptoms so you can remove infected growth quickly.

Fungi and bacteria both grow best in warm, damp conditions, so avoid overcrowding plants, ventilate your greenhouse, and don't prune after rainfall, when disease quickly spreads. Aphids and other sap-sucking insects carry viruses. Use barriers that keep pests away from plants to protect against infection (see pp.200–201). Dispose of diseased plants in local authority green waste, because spores may survive lower temperatures (see p.192) in your compost heap. Pathogens often only infect closely related plants, so keep plantings varied and rotate vegetable crops to limit spread (see p.95).

APHID ATTACK

Aphids can inject viruses while feeding

Thin leaf cuticle is easily pierced

Holes in leaf are gateways for infection

BACTERIAL DAMAGE

Microbes enter plant through leaf injury

Bacteria within leaf quickly cause disease

UNHEALTHY PLANT

Struggling plants have weak defences and are less able to repel infection. Injuries and pest damage are easy entry points for disease, and aphids may even transfer viruses directly. Roots in wet soil are also prone to attack.

Waterlogged soil weakens plant and promotes root rots

WHAT ARE THE DIFFERENT WAYS INSECTS ATTACK PLANTS?

Plants provide nourishment for most of the animal kingdom and insects and invertebrates are no exception. This becomes a problem for us when garden plants are on their menu, but you can piece together clues to deduce the likely culprit.

DON'T SUCK
SINCE SAP FLOWS UNDER PRESSURE, APHIDS HAVE NO NEED TO SUCK AND SIMPLY PIERCE THE PLANT SURFACE

Excess sap excreted as honeydew

Mouthparts tap into sap and control flow with muscular valve

Plant sap

We cannot eat grass because our intestines lack the digestive chemicals to break down cellulose, the main energy-containing substance in leaves and stems. Many common insects that leave bite marks on plants' leaves, stems, and flowers have no such problem as their guts contain cellulose-digesting enzymes. Small bite marks and holes in leaves come from small-mouthed creatures, such as leaf beetles, flea beetles, and young caterpillars. As caterpillars grow, so do their appetites, and their pincer-shaped jaws tear off larger pieces, leaving irregular notches around the edges of leaves or even stripping whole leaves so that only their central veins remain. Weevils and grasshoppers cause similarly large bites that are in proportion to their mouth size.

Woody stems present a challenge to potential predators. Their juicy innards are protected by a tough bark coat, which can only be penetrated by specialist "borer" insects with powerful muscles and strong jaws, such as bark beetles and the larvae (caterpillar form) of several moths.

Aphids colonize a rose bud Sap-sucking insects can often be found at the tips of new shoots, where soft growth is easy to pierce and full of sap.

TAPPING A PLANT'S SAP

Insects with simple innards cannot handle solid food and instead rely on plants' nutrient-filled sap. Termed "sap-suckers", these insects plunge their needle-like mouthparts (called a stylet) into leaves and soft stems to draw off sap flowing through the plant. Sap-suckers include aphids (greenfly and blackfly), scale insects, spider mites, and whiteflies, which cluster on stems, and under leaves and petals. This feeding can cause discoloration and deformity in new growth, can weaken infested plants, and may infect plants with viruses (see p.194). Leaves can also be covered in sticky honeydew, which is secreted by aphids and scale insects. Thrips are even smaller sapsuckers with short stylets capable of only drawing out juices from cells just beneath a leaf surface. Thrip damage is seen as tiny black or pale dots.

Not content to suck out a plant's life juices, the saliva of some insects and mites contains irritating chemicals, which can increase plant hormone levels in damaged leaves and cause a wart-like swelling called a gall. Insects often manipulate plant defences to produce these strange growths as an incubator and food source for their young, by injecting their eggs into a leaf or stem so that they are enveloped within the protective gall.

ROOTS AND FRUITS

A host of flying insects lay their eggs in the soil close to plants, so that the energy-packed roots, rhizomes, tubers, and bulbs, can provide sustenance for their larvae (also called maggots or grubs) when they hatch. Cabbage root fly larvae, chafer grubs, vine weevil grubs, and leatherjackets all damage plants in this way. Other root feeders include eelworms, root aphids, and root mealybugs. Gardeners only usually become aware of their presence when roots are so badly damaged that growth slows and plants wilt.

Fruits are laden with sugars and made to be eaten, although not by insects. Ripe berries and fruits are easy pickings, especially if their skin is soft or damaged, and are readily devoured by creepy crawlies big and small – from fruit flies to beetles, wasps, and earwigs. Developing fruit can fall victim to insect larvae with boring mouths, such as apple sawfly and codling moth, which tunnel into the core through a weak spot at the flower end of the fruit.

Vine weevil life cycle

Like many insects, these beetles feed on different parts of plants in their larval and adult phases.

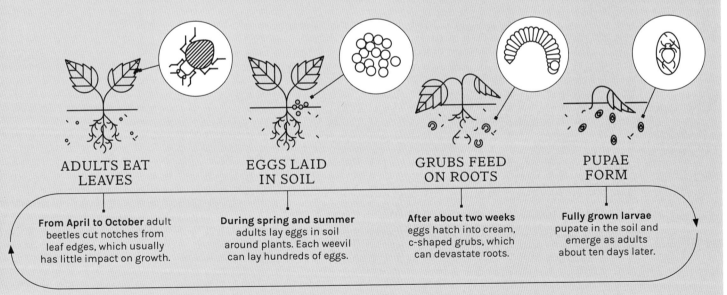

ADULTS EAT LEAVES

From April to October adult beetles cut notches from leaf edges, which usually has little impact on growth.

EGGS LAID IN SOIL

During spring and summer adults lay eggs in soil around plants. Each weevil can lay hundreds of eggs.

GRUBS FEED ON ROOTS

After about two weeks eggs hatch into cream, c-shaped grubs, which can devastate roots.

PUPAE FORM

Fully grown larvae pupate in the soil and emerge as adults about ten days later.

WHAT'S THE BEST WAY TO PREVENT PEST ATTACK?

Few things are as disheartening as chewed dahlias or cabbages shot with holes, but resist the urge to reach for a spray. Focusing on killing the pest misses the bigger picture and could worsen the situation for your plants and the wider environment.

Rather than look for a magic bullet, help plants fight their own corner by using a combination of tactics, tailored to the plants, likely pests, and conditions in your garden. This is called "integrated pest management" (see pp.52–53), and involves growing the strongest possible plants and taking steps to naturally control, deter, and exclude pests, so that insect-killing sprays are avoided.

WEAK PLANTS ARE EASY PICKINGS

Healthy people have fewer illnesses and infections, and the same is true of well-tended plants living in healthy soil (see pp.42–43). Abundant sugar and nutrients power their immune systems, allowing them to quickly detect and respond to pest damage. Within each plant cell is molecular machinery that detects a predator's saliva and can set off a rapid chain reaction which culminates in pest-repelling substances accumulating around the site of the bite. Tiny seedlings are extremely vulnerable to pests and

are often best started in ideal conditions under cover (see pp.82–83), to be transplanted when they have true leaves and can grow more rapidly.

CHOOSE PLANTS CAREFULLY

Making sure that you pick the right plant for the right place (see pp.58–59) will give plants the best start and maximize their pest repelling powers as they grow. Seek out plant species and varieties that are not targeted by the pests present in your area and avoid those that are vulnerable.

Research shows that gardens hosting a wide variety of plants have fewer insect pests than those with a limited selection. This is because a range of plants offers food sources for a variety of insects, many of which will be beneficial and prey on pests. "Companion planting" theory suggests that some plants repel pests, while others attract them and can be used as a "trap crop", but evidence to support these strategies is very hit and miss. Growing

TYPES OF TRAPS AND BARRIERS

A variety of traps and barriers is available to gardeners to help detect insect pests and keep them away from plants, but they will not control their numbers. Pheromones are chemical signals released by insects and other animals, here recreated synthetically to control pest numbers by luring them into traps or by disrupting their reproductive cycle.

PHEROMONE TRAPS

Detect:
Codling moth - apple
Plum moth - plum
Box moth - box (*Buxus*)

STICKY YELLOW TAPE

Detect:
Whitefly - greenhouse
Aphids - greenhouse
Thrips - greenhouse

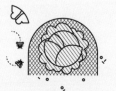

NETTING/INSECT MESH

Control:
Butterflies - brassicas
Carrot fly - carrots
Flea beetles - brassicas

See the bigger picture
Rather than targeting pests, try to view the garden as an ecosystem, where plants and pests are interlinked with each other and the growing conditions. A garden that is a well-functioning ecosystem is more environmentally sustainable and produces healthier plants.

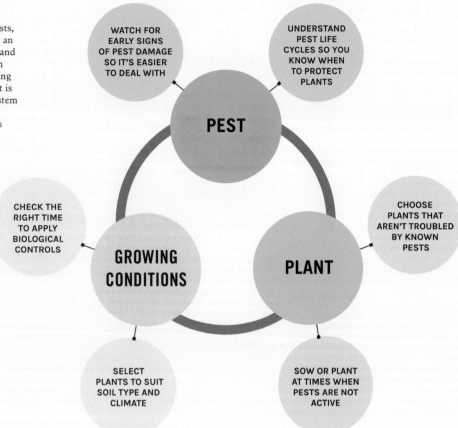

WATCH FOR EARLY SIGNS OF PEST DAMAGE SO IT'S EASIER TO DEAL WITH

UNDERSTAND PEST LIFE CYCLES SO YOU KNOW WHEN TO PROTECT PLANTS

PEST

CHECK THE RIGHT TIME TO APPLY BIOLOGICAL CONTROLS

GROWING CONDITIONS

PLANT

CHOOSE PLANTS THAT AREN'T TROUBLED BY KNOWN PESTS

SELECT PLANTS TO SUIT SOIL TYPE AND CLIMATE

SOW OR PLANT AT TIMES WHEN PESTS ARE NOT ACTIVE

different crop plants in a bed each year ("crop rotation") can also help to prevent problems with a handful of pests that overwinter in the soil and re-emerge in spring, but can be tricky to practice in small gardens with limited space.

TRAPS, DETERRENTS, AND BARRIERS

Traps are unlikely to offer complete protection – use them to monitor the number of insects in an area, rather than eradicate them. Only a proportion of a pest will be caught and traps baited with plant scents or pheromones may actually attract would-be attackers to the garden.

Physical barriers, such as fine insect mesh, can be invaluable for keeping flying insects and plants apart. They are extremely effective at keeping insect pests and birds away from fruit and vegetable

crops. A sprayed-on coating of kaolin clay solution (a powdery mineral that irritates insects) can act as a further deterrent.

BIOLOGICAL CONTROLS

Releasing natural pest predators ("biological" controls) can deal a fatal blow to pests, but most will only be active within certain temperature ranges and are often most effective in the enclosed environment of a greenhouse. Most beneficial insects released into a garden will probably die or move on to find food elsewhere. Chemical pesticides can be used if all else fails, but are likely to kill many helpful insects, including those sold as biological controls. Choose the least toxic and fastest degrading product to minimize harm to the beneficial wildlife that naturally controls pest populations (see pp.52–53).

HOW CAN I PREVENT SLUGS AND SNAILS DAMAGING PLANTS?

Slugs and snails are misunderstood: although they can decimate prized plants, most prefer to eat dead matter, and are an integral part of the soil food web (see pp.44–45). The challenge for gardeners is to keep plants safe, without harming wildlife.

In recognition of the important role they play in the wider life of the garden, slugs and snails are no longer classified as pests by the UK's Royal Horticultural Society. As well as helping to recycle dead plant material, they are also prey for garden wildlife, including beetles, birds, toads, frogs, snakes, and hedgehogs. Given that these animals can raze a bed of seedlings overnight (which is when they feed), it's sensible to encourage their predators into the garden to keep numbers in check, perhaps by making a pond. You can also protect plants by reducing dark, damp, or overgrown hiding places nearby, and raising seedlings undercover (see pp.82–83) to plant out when large enough to tolerate damage.

MANUAL CONTROL AND BARRIERS

Regularly picking up slugs and snails by torchlight on damp evenings is one way to help control numbers and protect plants. During the day, they shelter in shaded spots under plant pots, stones, and in dense vegetation, so avoid creating these havens near vulnerable plants and carry out a daytime trawl of likely hiding places. The question is then what to do with those that you find? Some gardeners are willing to kill them, but if they are to be spared, then release them well away from your garden, so that they don't return.

The most widely used deterrents are probably barriers of eggshells, coffee grounds, grit, pine needles, or wool pellets laid on soil around plants in the hope that slugs and snails will not cross. When tested scientifically, however, none of these barriers worked, as the animal's slimy foot simply slides over them. Copper barriers don't give them an electrical shock, as is sometimes claimed, but may act as a deterrent if wide enough. A thick line of

TRAPS AND BARRIERS are often employed by gardeners to keep slugs and snails at bay, but most offer limited, or even no protection to plants. Traps only capture a small proportion of a garden's population and risk attracting hungry slugs if placed near plants. Barriers are proven to be ineffective.

EGG SHELL BARRIER

Ineffective
Crushed shells and other sharp barriers won't stop these molluscs.

GRAPEFRUIT SKIN

Limited protection
Left face down on soil and emptied regularly, they help lower numbers.

COPPER BARRIER

Largely ineffective
Research has found that these only work if more than 4cm (1¹/₂in) wide.

BEER TRAP

Limited protection
Certainly attracts and kills molluscs, but will never catch them all.

MOVEMENT

THE SLIME-COATED MUSCULAR "FOOT" OF A SNAIL GLIDES OVER SHARP OR ROUGH SURFACES WITHOUT A SCRATCH

EATING

THE JAW CUTS OFF PLANT MATERIAL TO BE SCRAPED BY A RADULA OF THOUSANDS OF MICROSCOPIC TEETH

Snails and slugs are adapted to consume soft plant tissue, and contrary to myth their progress is not easily impeded.

diatomaceous earth – a powdered rock like microscopic shards of glass, which slugs and snails won't crawl across – does work, but the effect is lost if it gets wet.

SETTING TRAPS

"Beer traps" are simply small containers filled with beer, left buried in the ground overnight, with a lip protruding above the soil to prevent other creepy crawlies falling in. Slugs have poor eyesight but exquisitely sensitive scent-sensing tentacles and a peculiar penchant for the aroma of lager. Research that compared beer brands in traps found that, of the beers tested, Budweiser consistently caught the most slugs (fresh being better than flat). Lured by the promise of a tipple, a minority will lose their footing and slip into the drink and drown, but most simply slither away, meaning that beer traps can only help control the population rather than offer complete protection. The skin of half an orange or grapefruit can also be left on soil as a "trap" to draw slugs and snails in with the aroma of citrus. Lift

the skins in the morning and any residents can be removed, again helping to control the population, but providing limited protection.

PROVEN CONTROL METHODS

More effective for protecting plants from slugs are biological controls containing microscopic worms (called nematodes), which can be applied to soil, where they infect and kill slugs (but not snails) without harming wildlife. Pellets laced with a molluscicide (a chemical that kills slugs and snails) will certainly wipe them out, but traditional slug pellets contain metaldehyde, which is also toxic to pets, children, and wildlife. It readily washes into waterways, and has been detected in drinking water. Metaldehyde products were banned from sale and use in the UK in 2022, the first country to have done so. Alternative slug killers containing iron phosphate are less harmful to the environment. Research shows they are effective, although this is less obvious because slugs and snails aren't killed immediately and tend to die out of sight.

JARGON BUSTER

As beautifully simple as gardening can be, seasoned gardeners often use a lot of strange terms and funny-sounding words, which can make it complicated to work out where to start. Untangling this terminology is a great way to improve your understanding and bring on your skills.

ANNUAL

PLANTS THAT GERMINATE, GROW, REPRODUCE, AND DIE WITHIN ONE CALENDAR YEAR. THIS LIVE FAST AND DIE YOUNG STRATEGY EVOLVED TO TAKE ADVANTAGE OF SHORT GROWING SEASONS IN INHOSPITABLE PLACES.

BIENNIAL

BIENNIALS (FROM "EVERY TWO YEARS") FLOWER IN THEIR SECOND GROWING SEASON, AND THEN DIE. THE FIRST GROWING SEASON IS USED TO BUILD UP THE RESOURCES NEEDED TO FLOWER AND PRODUCE SEEDS.

PERENNIAL

AROUND 90 PER CENT OF ALL PLANTS ARE PERENNIALS (MEANING "THROUGH THE YEAR"), WHICH DO NOT DIE AFTER FLOWERING AND CAN LIVE FOR MANY YEARS, SOMETIMES INDEFINITELY.

SHRUB

A WOODY PLANT WITH SEVERAL MAIN STEMS GROWING FROM GROUND LEVEL. CAN SOMETIMES BE PRUNED AND TRAINED TO LOOK LIKE A TREE.

TREE

A WOODY PLANT WITH A SINGLE STEM, CALLED A TRUNK, FROM WHICH BRANCHES GROW OUTWARDS TO FORM A BUSHY HEAD, KNOWN AS A CROWN. MAY ALSO OCCASIONALLY BE MULTI-STEMMED.

WOODY

PLANTS WITH STIFF, BARK-COVERED STEMS WHICH SUPPORT NEW GROWTH YEAR AFTER YEAR. WOOD IS MADE FROM COMPRESSED PLANT FIBRES (CELLULOSE) HELD TOGETHER BY A POWERFUL GLUE (LIGNIN). ALL WOODY PLANTS ARE PERENNIAL.

DECIDUOUS

PERENNIAL PLANTS THAT DELIBERATELY DUMP THEIR LEAVES IN AUTUMN TO REDUCE ENERGY LOSS AND THE RISK OF DAMAGE TO THE PLANT DURING WINTER (SEE PP.154–155).

EVERGREEN

PLANTS THAT KEEP THEIR LEAVES THROUGHOUT THE YEAR. NOT TO BE CONFUSED WITH CONIFERS, WHICH ARE CONE-BEARING PLANTS THAT SOMETIMES LOSE THEIR LEAVES.

HERBACEOUS

SOFT-STEMMED PLANTS, WITH NO WOODY TISSUE. BECAUSE WOOD TAKES TIME TO BUILD, ALL ANNUALS AND BIENNIALS ARE HERBACEOUS, BUT MANY PERENNIALS ARE TOO.

Legend

- Plant life cycles
- Plant stem structure
- Plant forms
- Leaf growth cycles
- Practical terms – young plants
- Practical terms – shaping growth

POT ON

MOVING A POTTED PLANT INTO A LARGER CONTAINER, ALSO CALLED "POTTING UP". THIS COULD BE TRANSFERRING A SEEDLING FROM A TRAY OF INDIVIDUAL MODULES INTO A SMALL POT, OR A HOUSEPLANT INTO A BIGGER POT.

PRICK OUT

TO TRANSFER A SMALL SEEDLING FROM WHERE IT WAS SOWN (USUALLY A SEED TRAY) TO A NEW CONTAINER. THIS GIVES YOUNG PLANTS MORE RESOURCES, SPACE, AND LIGHT TO GROW STRONGLY.

TRAIN

TO FORCE A PLANT TO GROW IN A PARTICULAR DIRECTION OR SHAPE BY POSITIONING STEMS WHILE YOUNG AND BENDY; USUALLY BY TYING STEMS TO HORIZONTAL WIRES, TRELLIS, OR OTHER SUPPORTS.

HARDEN OFF

READYING AN INDOOR-RAISED PLANT TO BE MOVED OUTSIDE, BY LEAVING IT OUTDOORS FOR PROGRESSIVELY LONGER PERIODS. THIS ACTIVATES ITS BIOLOGICAL DEFENCES TO WIND, SUN, AND COLD.

PRUNE

TO DELIBERATELY CUT OFF ANY PLANT PART WITH A VIEW TO DIVERTING PLANT GROWTH IN A CHOSEN DIRECTION, CHANGING A PLANT'S FORM OR IMPROVING FLOWERING OR FRUITING (SEE PP.164–165).

DEADHEAD

THE DECAPITATION OF WILTING FLOWERS BEFORE SEEDS FORM, IN AN EFFORT TO REDIRECT A PLANT'S ENERGY INTO PRODUCING MORE FLOWERS.

TRANSPLANT

THE ACTION OF MOVING A PLANT TO A NEW HOME, WHICH COULD BE FROM A POT INTO GARDEN SOIL OR FROM ONE PLACE IN THE GARDEN TO ANOTHER.

THIN OUT

REMOVING YOUNG PLANTS (USUALLY SEEDLINGS) FROM OVERCROWDED ROWS OR POTS, SO THAT THE REMAINING PLANTS HAVE SPACE TO THRIVE.

PINCH OUT

TO NIP OFF A SHOOT TIP, BETWEEN THUMB AND FOREFINGER TO ENCOURAGE BUSHY GROWTH. OR TO REMOVE A SIDESHOOT TO PROMOTE UPWARD GROWTH.

BOLTING

WHEN A PLANT GROWN FOR ITS EDIBLE LEAVES, ROOTS, OR BULB SENDS UP A FLOWERING STEM, DRAINING THE EDIBLE PARTS OF RESOURCES AND OFTEN MAKING THEM SMALL AND BITTER. (SEE PP.146–147).

NECTAR

A SUGARY LIQUID (UP TO 80 PER CENT SUGAR) THAT IS RELEASED IN TINY AMOUNTS IN FLOWERS TO LURE POTENTIAL POLLINATORS. PRODUCED IN NECTARY GLANDS, USUALLY HIDDEN NEAR THE BASE OF THE FLOWER.

CELLS

MICROSCOPICALLY SMALL LIVING BUBBLES THAT MAKE UP ALL PLANTS AND ANIMALS. EACH CONTAINS THE BIOLOGICAL MACHINERY OF LIFE, ALONG WITH DNA, WHICH PROVIDES THE OPERATING INSTRUCTIONS. (SEE PP.72–73).

POLLINATION

EQUIVALENT TO PLANT SEX: THE COMING TOGETHER OF MALE POLLEN, WITH THE FEMALE REPRODUCTIVE PARTS OF A FLOWER THAT HOUSE THE EGG.

POLLEN

THE PLANT EQUIVALENT OF SPERM. EACH MICROSCOPIC GRAIN IS A CELL CONTAINING MALE DNA. POLLEN IS HELD ALOFT IN BUNDLES (ANTHERS) ON TALL STALKS WITHIN FLOWERS, (STAMEN).

POLLINATOR

ANY ANIMAL THAT PERFORMS REPRODUCTIVE SERVICES FOR PLANTS, BY CARRYING STICKY POLLEN GRAINS FROM ONE FLOWER TO ANOTHER.

ORGANIC MATTER

DEAD, DECAYING, OR DECOMPOSED MATERIAL THAT HAS COME FROM A LIVING PLANT OR ANIMAL; MAKES UP A SMALL BUT VITAL PART OF HEALTHY SOIL (SEE PP.42–43).

COMPOST

DEAD PLANT OR ANIMAL MATERIAL (ORGANIC MATTER) THAT HAS BEEN DECOMPOSED BY BACTERIA, FUNGI, AND OTHER CREATURES; ALSO USED TO REFER TO POTTING MIXES (SEE PP.188–189).

MULCH

ANY SUBSTANCE LAID OVER SOIL TO COVER ITS SURFACE. ALL MULCHES PREVENT MOISTURE EVAPORATING FROM SOIL AND SUPRESS WEED GROWTH. MULCHES OF ORGANIC MATTER ALSO "FEED" SOIL (SEE PP.42–43).

Flowers and reproduction

Cells

Organic gardening

Soil health and improvement

First growth

Temperature sensitivity

GERMINATION

THE SPROUTING OF A SEED, WHICH BEGINS WHEN A SEED SWELLS WITH WATER AND ENDS WHEN THE BABY ROOT ("RADICLE") EMERGES (SEE PP.76–77).

SEED LEAVES

KNOWN AS "COTYLEDONS", THESE ARE FIRST LEAVES TO EMERGE FROM THE SEED, WHICH WERE FOLDED UP INSIDE, AND ARE FILLED WITH FUEL FOR THE FIRST FEW DAYS OF LIFE.

TRUE LEAVES

FULLY FORMED LEAVES WITH THE APPEARANCE AND CHARACTERISTICS OF THE MATURE PLANT. THEY APPEAR AFTER THE SEED LEAVES, AS A SEEDLING GROWS.

ORGANIC GARDENING

A GARDENING PHILOSOPHY THAT EMPHASIZES "WORKING WITH NATURE" BY IMPROVING SOIL HEALTH AND AVOIDING THE USE OF SYNTHETIC FERTILIZERS AND PESTICIDES (SEE PP.54–55).

TENDER

DESCRIBES PLANTS THAT WILL DIE IF EXPOSED TO FROST OR COLD WINTER TEMPERATURES, AND WOULD NEED TO BE PROTECTED OR MOVED UNDER COVER TO SURVIVE WINTER IN COLDER CLIMATES (SEE PP.80–81).

HARDY

ANY PLANTS ABLE TO SURVIVE WINTER OUTDOORS, THANKS TO A VARIETY OF BIOLOGICAL PROTECTIONS AGAINST LOW TEMPERATURES (SEE PP.80–81).

MANURE

ANIMAL WASTE THAT IS APPLIED TO SOIL TO IMPROVE ITS HEALTH AND ADD NUTRIENTS; FRESH MANURE CAN CONTAIN HARMFUL BACTERIA AND HIGH LEVELS OF AMMONIA, AND IS BEST LEFT TO DECOMPOSE (ROT) BEFORE USING (SEE P.193).

GREEN MANURE

FAST-GROWING "COVER CROP" PLANTS SOWN INTO BARE SOIL TO PREVENT EROSION AND WEED GROWTH; ALSO NOURISHES SOIL WHEN IT IS DUG IN OR UPROOTED AND LEFT TO DECOMPOSE (SEE PP.42–43).

SOIL FOOD WEB

THE COUNTLESS ORGANISMS, VISIBLE AND MICROSCOPIC, WHICH LIVE IN OR ON THE SOIL, RECYCLING NUTRIENTS, IMPROVING SOIL STRUCTURE, AND SUPPORTING PLANT GROWTH (SEE PP.44–45).

GARDENING MYTHS

Many traditional practices and more modern ideas have entered into gardening lore. But research has separated myth from fact to reveal that some of these, often time consuming, methods have little benefit or may even be harmful.

———

EGG SHELLS AND BEER TRAPS STOP SLUG AND SNAIL DAMAGE

THE SHARP EDGES OF EGG SHELLS ARE NO BARRIER TO THESE CREATURES.

Their muscular, slime-covered feet will slide over barriers to reach a tempting plant. Beer traps capture only the handful of slugs unfortunate enough to slip in, and so don't prevent damage to plants.

SEE PP.210–211

MISTING HELPS INDOOR PLANTS GROW

PLANTS ARE OFTEN SPRAYED WITH A FINE MIST OF WATER TO INCREASE HUMIDITY.

But this only increases humidity around leaves very fleetingly and is of little benefit in the dry air of a modern house.

SEE PP.120–121

HOUSEPLANTS NEED A WATERING SCHEDULE

OVERWATERING IS THE NUMBER ONE CAUSE OF DEATH FOR INDOOR PLANTS.

Forget the watering reminder apps and instead water plants according to their changing needs through the seasons.

SEE PP.110–113, P.163

WATERING UNDER HOT SUN BURNS LEAVES

GARDENING ADVICE IS TO AVOID WATERING PLANTS UNDER BLAZING SUMMER SUN.

It is said that water droplets focus solar rays into the leaves, causing burning or scorching. This "lens effect" never happens, not least because droplets evaporate away far too quickly. If plants are thirsty, give them water.

SEE PP.114–115

PLANTS HELP CLEAN INDOOR AIR, INCREASING OXYGEN LEVELS

PLANTS DO RELEASE OXYGEN AS THEY PRODUCE FOOD FROM PHOTOSYNTHESIS.

But a houseplant produces less than a thousandth of the oxygen that you breathe every day. Plants also absorb some air pollution, but when you do the sums, the hundreds of plants needed to meaningfully remove harmful vapours in your home would likely leave you with little space to stand.

SEE P.20, PP.70-71

PRUNING CUTS SHOULD BE MADE AT AN ANGLE

FLAT CUTS ARE ACTUALLY THE SMALLEST AND FASTEST HEALING PRUNING CUTS.

Angled cuts leave a larger wound, take longer to heal, and do not prevent rot by stopping water pooling on the stem.

SEE PP.170-171

WATER EVAPORATING FROM A PEBBLE-FILLED SAUCER RAISES HUMIDITY

PUTTING PLANTS ON A DISH OF WET PEBBLES IS SAID TO RAISE HUMIDITY AROUND LEAVES.

This may look pretty, but does nothing to change the humidity in the air around their foliage.

SEE PP.120-121

ORGANIC PESTICIDES ARE SAFER THAN SYNTHETIC VERSIONS

ALL PESTICIDES CAN KILL BENEFICIAL WILDLIFE AS WELL AS PESTS, SO NONE CAN BE USED WITHOUT RISKING HARM.

For many good reasons, synthetic pesticides get a bad rap. Also known as "conventional" pesticides, the most notorious of these manufactured chemicals have been linked to cancer, Alzheimer's disease, ADHD, and even birth defects, and they all break down slowly and linger in soil for months or years. Organic pesticides are also poisons, just extracted from plants or concocted out of minerals in a lab, although this does mean they tend to break down faster in the environment and cause less long-term damage.

SEE PP.52-53

ADDING SAND TO CLAY SOIL IMPROVES DRAINAGE

HEAVY CLAY SOILS ARE OFTEN WATERLOGGED WHEN WET, MAKING GARDENING DIFFICULT.

It's no wonder then that gardeners have long been trying to ease water's flow through clay soils by digging in sand or grit. It makes perfect sense: water drains easily through sandy soils and so increasing the proportion of sand in clay soil should make it better draining. In reality, it's not practical to add enough sand or grit to offset clay's sticky properties. Light soils are at least fifty per cent sand, and you would need to dig in around 250kg (550lb) for every square metre of clay soil to achieve this composition. Furthermore, any digging will make drainage worse by destroying soil's structure of tiny pores and slowing water's flow through it.

SEE PP.38–39, PP.42–43

ADDING LIME MAKES SOIL LESS ACIDIC

LIME (LIMESTONE) DOES NEUTRALIZE ACIDS IN SOIL, MAKING IT LESS ACIDIC.

This raises the soil's pH. However, the minerals and organic matter within soil naturally resist or "buffer" any enforced change and, even with an accurate pH reading, it is impossible to know how much lime is needed to overcome a soil's buffering. Regardless, soil will gradually revert to its original pH anyway.

SEE PP.40–41

A LAYER OF CROCKS OR STONES AT THE BOTTOM OF CONTAINERS IMPROVES DRAINAGE

PLANTS IN WATERLOGGED SOIL QUICKLY FALL VICTIM TO FATAL FUNGAL ROOT ROTS.

The popular advice to prevent this is to place pieces of broken pots (termed "crocks") or gravel into the pot before topping up with potting compost. However, science shows that plants in containers with crocks fare no better than those without – in fact, their addition may actually prevent good drainage. The small pores between soil particles hold onto water like a sponge, so that it does not easily flow into the much larger spaces between crocks or gravel. Instead, water clings to the lowest layer of soil, where it can accumulate and cause drainage problems. The best advice to avoid waterlogging is to use good quality potting mix, a pot with drainage holes, and to not overwater.

SEE PP.110–111

ONLY NATIVE PLANTS FEED BENEFICIAL INSECTS

FLOWERING PLANTS AND INSECTS EVOLVED LIKE A COUPLE DANCING THE TANGO:

Flowers tweaked their proportions, scents, and colours to increase their appeal to local bugs, while insects feeding from particular flowers evolved their anatomy to gather pollen and nectar more effectively. These relationships mean that today, some insects are only able to feed on a narrow range of plants. Growing plants from your region ("natives") will likely best service such specialist pollinators, but most insects are not limited in this way and science actually shows that gardens with a variety of native and non-native flowers support the most wildlife.

SEE PP.62–63, PP.140–141

TALKING TO PLANTS BENEFITS THEIR GROWTH

PRINCE CHARLES HAS LONG BEEN RIDICULED FOR CLAIMING THAT TALKING TO PLANTS HELPS THEM GROW.

As silly as this sounds, science shows that plants feel the air vibrations that cause sound, and grow faster when placed in front of speakers playing a continuous tone or music. This is possibly because they have evolved to sense the wind and contact from animals and insects, and this stimulation is a natural part of their development. It seems unlikely that occasional words would boost growth, although it has never been conclusively proved one way or the other. Human breath also contains high levels of carbon dioxide, which plants use to make food via photosynthesis, but whether this fleeting increase affects growth is unknown.

SEE PP.70–71, PP.86–87

URBAN BEEKEEPING HELPS HONEYBEES AND OTHER POLLINATORS

INSECT POLLINATORS ARE BEING DECIMATED BY CLIMATE CHANGE, HABITAT DESTRUCTION, AND PESTICIDE OVERUSE.

Although siting beehives in urban areas is promoted as a solution, this brings its own problems. A new beehive introduces a ravenous colony of honeybees to gobble up all the nearby nectar and pollen, leaving the local pollinators to starve, which includes bumblebees, solitary bees, hoverflies, moths, and butterflies. Research shows that in areas with urban beehives, the numbers of other pollinators dwindle. Worse still, because there isn't enough food to go round, these beehives rarely produce a good crop of honey and are often abandoned after the damage has been done.

SEE PP.30–31

INDEX

A

abscisic acid 114, 117
abscission zone 154
acclimation 81
acid-loving plants 40, 41, 49, 110, 115
acidity, soil *see* pH
actinomycetes 189
aerenchyma 116
Agapanthus 59, 161
air plants (*Tillandsia*) 10, 108
alfalfa 95
alkaline soils 40
alliums 59, 105
Amaranthus 76
amaryllis (*Hippeastrum*) 99, 100
Anemone 59, 97
annual plants 57, 92, 144, 204
anthocyanins 154
aphids 17, 54, 94, 197, 198, 199
apical dominance 166–167, 170, 172
apple blossom 24
apple sawfly 199
aspect 32–33, 36, 58
autumn leaf colour 154–155
auxin 107, 137, 166, 172, 173, 184

B

bacteria, soil 14, 30, 40, 44, 45, 159, 188, 189, 190
bacterial diseases 169, 194, 195, 197
balconies 18
bamboos 90, 97
Banksia 77
basic needs of plants 70–71

beekeeping 211
beer traps 202, 203, 211
bees 94, 134, 138, 140, 141
beetroot 92, 177
biennial bearing 149
biennials 78, 147, 204
binomial nomenclature 64–65
birch (*Betula*) 169
black-eyed Susan (*Rudbeckia fulgida*) 141, 182
blackberries 128, 186
blackcurrants (*Ribes nigrum*) 165, 186
blight 196
bolting 146–147, 206
bonsai 92
borage (*Borago officinale*) 141
boron 122, 123
Boston ivy (*Parthenocissus tricuspidata*) 128
botanical groups 64–65
box blight 197
breeding new varieties 180–181
breeding true 180
bristlecone pine (*Pinus longaeva*) 10
broad beans 74, 177
broccoli 35
Brunsvigia 11
bugle (*Ajuga reptans*) 59, 186
bulbs 96–99
    bulblets 187
    forcing 99
    knotting the foliage 156
    lifting 157, 187
    planting depth 98
    planting "in the green" 78
    propagation 137, 187

"right way up" 98
    scales 97, 187
    shoots 99
    spring-flowering 99
    see also corms; rhizomes; tubers
busy lizzy (*Impatiens*) 142, 197
buttercup (*Ranunculus*) 64
butterfly bush (*Buddleja davidii*) 148, 165, 169
buying plants 18–19
    bare root plants 78

C

cabbage root fly 199
cacti 100, 101, 117, 142, 143
Caladium 97
calcium 40, 115, 122, 123, 124
calluses 90, 91, 106, 167, 182–183
cambiums 90, 91
Camellia 118
canker 169, 194
Canna 97, 157
cape primrose (*Streptocarpus*) 142
carbon cycle 21
carbon dioxide 18, 30, 51, 70, 71, 159, 208
    carbon capture and storage 19, 21
    and photosynthesis 70, 117
carnivorous plants 108
carpetweed (*Phyla canescens*) 63
carrots 82–83, 147
cast iron plant (*Aspidistra elatior*) 142
caterpillars 63, 198
catmint (*Nepeta*) 59, 159
cells 206

plant cells 72–73
stem cells 90, 136–137, 175, 184
cellulose 72, 198, 204
Celosia 144
Chaenomeles 169
chafer grubs 199
Chinese juniper (*Juniperus chinensis*) 20
chipping 187
chlorine 122, 123
chlorophyll 70, 122, 154, 175, 194
chloroplasts 35, 70, 71, 72
chlorosis 40, 122, 194
    interveinal chlorosis 194
Christmas cactus (*Schlumbergera*) 142
circadian clock 142
circular economy 31
circumnutation 128
Citrus 35, 142–143
clay soils 39, 58, 75, 161
    drainage 40, 210
    pH 40
    water retention 23, 39, 113
Clematis 128, 131, 165, 171
climate
    local 34–35
    microclimates 36–37, 58
climate change, gardens and 18, 24–25
climbing plants
    climbing methods 13, 128–129
    cuttings 184
    scandent plants 129
    self-clinging climbers 131
    spacing 92
    supports 131
cloches 161
clover (*Trifolium*) 42, 60, 61, 124
coco de mer 74

218

BIBLIOGRAPHY

10–11 HOW AMAZING ARE PLANTS?
Krishna Kumar Kandaswamy et al., "AFP-Pred: A random forest approach for predicting antifreeze proteins from sequence-derived properties", *Journal of Theoretical Biology* 270 (2011) 56–62. Matt Candeias, *In Defense of Plants: an exploration into the wonder of plants*, Mango, 2021.

12–13 DO PLANTS HAVE INTELLIGENCE?
Stefano Mancuso, *The Roots of Plant Intelligence*, ted.com/talks/ stefano_mancuso_the_roots_of_plant_intelligence. F. Baluska et al., "The "root-brain" hypothesis of Charles and Francis Darwin: Revival after more than 125 years', *Plant Signal Behav.*, 4 (2009) 1121–1127. Michel Thellier et al., "Long-distance transport, storage and recall of morphogenetic information in plants. The existence of

a sort of primitive plant memory", *Comptes Rendus de l'Académie des Sciences - Series III - Sciences de la Vie*, 323 (2000), 81–91. H.M. Appel et al., "Plants respond to leaf vibrations caused by insect herbivore chewing", *Oecologia* 175 (2014) 1257–1266. M. Gagliano et al., "Tuned in: plant roots use sound to locate water", *Oecologia* 184 (2017) 151–160.

16–17 HOW IMPORTANT ARE GARDENS FOR WILDLIFE?
"Living Planet Report", WWF [web article], 2020, livingplanet. panda.org/en-us/. A. Derby Lewis et al., "Does Nature Need Cities? Pollinators Reveal a Role for Cities in Wildlife Conservation", *Frontiers in Ecology and Evolution*, 7 (2019). James G. Rodger et al., "Widespread vulnerability of flowering plant seed production to pollinator declines", *Science Advances,* 7 no. 42 (2021).

20 DO PLANTS ABSORB POLLUTANTS?
"9 out of 10 people worldwide breathe polluted air, but more countries are taking action", World Health Organisation [web article], 2 May 2018, who.int/news/item/02-05-2018-9-out-of-10-people-worldwide-breathe-polluted-air-but-more-countries-are-taking-action. Michel Le Page, "Does air pollution really kill nearly 9 million people each year?", *New Scientist*, 12 March 2019. Zhang Jiangli et al., "Improving Air Quality by Nitric Oxide Consumption of Climate-Resilient Trees Suitable for Urban Greening", *Frontiers in Plant Science*, 11 (2020). K. Wróblewska et al., "Effectiveness of plants and green infrastructure utilization in ambient particulate matter removal", *Environ. Sci. Eur.* 33, 110 (2021). A. Diener, P. Mudu, "How can vegetation protect us from air pollution? A critical review on green spaces' mitigation abilities for air-borne particles from a public health perspective – with implications for urban planning", *Science of The Total Environment* 796, (2021). Udeshika Weerakkody et al., "Quantification of the traffic-generated particulate matter capture by plant species in a living wall and evaluation of the important leaf characteristics", *Science of The Total Environment* 635 (2018). B. C. Wolverton et al., "Interior landscape plants for indoor pollution abatement", NASA September 15 1989.

21 CAN MY GARDEN SOIL CAPTURE CARBON DIOXIDE?
"How many trees needed to offset your carbon emissions?", Samson Opanda, [web article] 8billiontrees.com/carbon-offsets-credits/reduce-co2-emissions/how-many-trees-offset-carbon-emissions/. "How much CO2 does a tree absorb?", Viessman [web article], viessmann.co.uk/heating-advice/how-much-co2-does-tree-absorb.

22 CAN PLANTS KEEP YOU COOL?
M. A. Rahman, A.R. Ennos, "What we know and don't know about the cooling benefits of urban trees", *Trees and Design Action Group*, (2016). K. K. Y. Liu, B. Bass, "Performance of green roof systems", *Cool Roofing Symposium*, Atlanta, GA., May 12–13 2005. Ying-Ming Su, Chia-Hi Lin, "Removal of Indoor Carbon Dioxide and Formaldehyde Using Green Walls by Bird Nest Fern", *The Horticulture Journal*, 84, no.1 (2015) 69–76.

23 COULD MY GARDEN HELP PREVENT FLOODING?
"Extreme weather events in Europe", European Academies' Science Advisory Council [web article], 2018, easac.eu/fileadmin/PDF_s/reports_statements/Extreme_Weather/EASAC_Extreme_Weather_2018_web_23March.pdf. "Rain gardens", Royal Horticultural Society [web article], www.rhs.org.uk/garden-featuresrain-gardens.

24-25 HOW WILL CLIMATE CHANGE AFFECT MY GARDEN?
Bob Oakes, Yasmin Amer, "How Thoreau helped make Walden pond one of the best places to study climate change in the US", WBUR [web article], wbur.org/news/ 2017/07/12/studying-climate-change-walden-pond. U. Büntgen et al., "Plants in the UK flower a month earlier under recent warming", Proc. R. Soc. 289, no.1968 (2022). "Status of spring 2022", USA National Phenology Network [web article], usanpn.org/news/spring. D. Graczyk, M. Szwed, "Changes in the occurrence of late spring frost in Poland", *Agronomy* 10, no. 1835 (2020). I. W. Park, T. Ramirez-Parada, S. J. Mazer, "Advancing frost dates have reduced frost risk among most North American angiosperms since 1980." Global Change Biology, 27(1), 2020, 165–176.

34-35 HOW DOES THE WEATHER AFFECT WHAT I CAN GROW?
Jerry L. Hatfield, John H. Prueger, "Temperature extremes: effect on plant growth and development", *Weather and Climate Extremes*, 10, part A (2015) 4–10. Hu Shanshan et al., "Sensitivity and responses of chloroplasts to heat stress in plants", *Frontiers in Plant Science*, 11 (2020). "Citrus", Royal Horticultural Society [web article], rhs.org.uk/fruit/citrus/grow-your-own.

36-37 WHAT IS A MICROCLIMATE?
Maraveas, C. "Design of Tall Cable-Supported Windbreak Panels." *Open Journal of Civil Engineering*, 9 (2019), 106–122.

40-41 WHAT'S SOIL PH AND HOW DOES IT AFFECT WHAT I CAN GROW?
S. Singh et al., "Soil properties change earthworm diversity indices in different agro-ecosystem", *BMC Ecology*, 20, no. 27 (2020). "Soil pH: what it means", Donald Bickelhaupt, SUNY College of Environmental Science and Forestry [web article], esf.edu/pubprog/brochure/soilph/soilph.htm.

42-43 HOW CAN I BEST IMPROVE MY SOIL?
"The soil is alive", European Commission Convention on Biological Diversity (2008). P. Bonfante, A. Genre, "Mechanisms underlying beneficial plant–fungus interactions in mycorrhizal symbiosis", *Nat. Commun.* 1, no. 48 (2010). H. Bücking et al., "The role of the mycorrhizal symbiosis in nutrient uptake of plants and the regulatory mechanisms underlying these transport processes", in *Plant Science*, IntechOpen, 2012.

44-45 WHAT IS THE SOIL FOOD WEB?
N. J. Balfour, F. L. W. Ratnieks, "The disproportionate value of "weeds" to pollinators and biodiversity", *Journal of Applied Ecology*, 59, no. 5, (2022) 1209–1218. J. M. Baskin, C. C. Baskin, "Does seed dormancy play a role in the germination ecology of *Rumex crispus*?" *Weed Science*, 33, no. 3 (1985) 340–343. "Invasive non-native plants", Royal Horticultural Society [web article], rhs.org.uk/prevention-protection/invasive-non-native-plants. "The impact of glyphosate on soil health", Soil Association [web article], soilassociation.org/media/7202/glyphosate-and-soil-health-full-report.pdf.

46-47 SHOULD I GET RID OF WEEDS?
"Never let'em set seed", Robert Norris, Weed Science Society of America [web article], wssa.net/wssa/weed/articles/wssa-neverletemsetseed/.

49 WHAT TYPE OF COMPOST SHOULD I BUY?
"On Test: compost for raising plants", Gardening Which? [web article], 2022, which.co.uk/reviews/compost/article/best-compost-ahUv44C6lrR5.

50–51 WHAT'S SO BAD ABOUT USING PEAT?
Fereidoun Rezanezhad et al., "Structure of peat soils and implications for water storage, flow and solute transport: A review update for geochemists", *Chemical Geology*, 429 (2016) 75–84. "Garden", Pesticide Action Network North America [web article], panna.org/starting-home/garden.

52–53 DO I NEED TO USE PESTICIDES?
"Homeowner's guide to protecting frogs – lawn and garden care", US Fish and Wildlife Service [web article], 2000, dwr.virginia.gov/wp-content/uploads/homeowners-guide-frogs.pdf. Christie Wilcox, "Myth Busting 101: organic farming>conventional agriculture", Scientific American [web article], 2011, blogs.scientificamerican.com/science-sushi/httpblogsscientificamericancomscience-sushi20110718mythbusting-101-organic-farming-conventional-agriculture/. "Monograph on Glyphosate" WHO International Agency for Research on Cancer [web article], 2015, iarc.who.int/featured-news/media-centre-iarc-news-glyphosate/. A. H. C. van Bruggen, et al., "Indirect effects of the herbicide glyphosate on plant, animal and human health through its effects on microbial communities", *Front. Environ. Sci.*, 18 (2021).

56–57 WHAT MAKES A GARDEN LOWER MAINTENANCE?
"3 million front gardens have been completely paved since 2005. Let's try to reverse this trend", Hayley Monkton, Lowimpact [web article], 2015, lowimpact.org/posts/3-million-front-gardens-have-been-completely-paved-since-2005-lets-try-to-reverse-this-trend.

60–61 SHOULD I HAVE A LAWN?
"Water Calculator", Eco Lawn, www.eco-lawn.com. "Grass lawns are an ecological catastrophe", Lenore Hitchler, Only Natural Energy [web article], 2018, onlynaturalenergy.com/grass-lawns-are-an-ecological-catastrophe/.

62–63 SHOULD I GROW ONLY NATIVE PLANTS
Matthew L. Forister, et al., "The global distribution of diet breadth in insect herbivores", *PNAS*, 112, no. 2 (2014) 442–447. Chris D. Thomas, *Inheritors of the Earth: How Nature Is Thriving in an Age of Extinction*, Penguin, 2018. "Native and non-native plants for pollinators", Royal Horticultural Society [web article], rhs.org.uk/wildlife/native-and-non-native-plants-for-pollinators.

64–65 WHAT'S WITH ALL THIS LATIN?
Anna Pavord, *The Naming of Names*, Bloomsbury, 2007.

66–67 ARE SOME GARDENING TOOLS BETTER THAN OTHERS?
"Anvil or Bypass Secateurs", Robert Pavlis, Garden Myths [web article], www.garden myths.com/anvil-bypass-secateurs-pruners/.

70–71 WHAT DO PLANTS NEED TO GROW?
"Plants release more carbon dioxide into atmosphere than expected", Australian Nat. Uni. [web article], 2017, anu.edu.au/news/all-news/plants-release-more-carbon-dioxide-into-atmosphere-than-expected.

72–73 HOW DO PLANT CELLS FUNCTION?
F. W. Telewski, "Mechanosensing and plant growth regulators elicited during the thigmomorphogenetic response", *Frontiers in Forests and Global Change* 18 (2021).

74 WHAT IS A SEED?
J. Shen-Miller et al., "Exceptional seed longevity and robust growth: ancient sacred lotus from China", *American Journal of Botany* 82 (1995) 1367–1380.

76–77 WHAT DO SEEDS NEED TO GERMINATE?
W. Aufhammer et al., "Germination of grain amaranth (*Amaranthus hypochondriacus* × *A. hybridus*): effects of seed quality, temperature, light, and pesticides", *European Journal of Agronomy,* 8 (1998) 127–135. "Start seeds indoors: digging deeper, pt 3" Joe Lamp'l [web article], Feb 15 2018, joegardener.com/podcast/seed-starting-part-3/

78–79 WHY DO WE SOW AND PLANT AT DIFFERENT TIMES OF YEAR?
"Seed-sowing techniques", Royal Horticultural Society [web article], rhs.org.uk/advice/beginners-guide/vegetable-basics/seed-sowing-techniques. Christian Körner, "Winter crop growth at low temperature may hold the answer for alpine treeline formation", *Plant Ecology & Diversity*, 1, no.1, (2008) 3–11.

80–81 WHAT IS MEANT BY "HARDINESS" AND HOW IS IT MEASURED?
Victoria Wyckelsma, Peter John Houweling, "Your genetics influence how resilient you are to cold temperatures – new research", The Conversation [web article], February 25 2021, theconversation.com/your-genetics-influence-how-resilient-you-are-to-cold-temperatures-new-research-155975. *RHS Plant Finder 2013*, Royal Horticultural Society, 2013. USDA Plant Hardiness Zone Map, planthardiness.ars.usda.gov. *The European Garden Flora 2nd Edition*, James Cullen, Sabina G. Knees, H. Suzanne Cubey (eds), Cambridge, 2011.

84–85 HOW CAN I KEEP SEEDLING GROWTH STRONG?
Hendrik Poorter et al., "Pot size matters: A meta-analysis of the effects of rooting volume on plant growth", *Functional Plant Biology*, 39 (2012) 839–850. "Potting up: which pot size is correct for potting up?", Robert Pavlis, Garden Myths [web article], gardenmyths.com/potting-up-correct-pot-size/.

86–87 WHAT IS "HARDENING OFF"?
E. Wassim Chehab et al., "Thigmomorphogenesis: a complex plant response to mechano-stimulation", *Journal of Experimental Botany*, 60, no. 1 (2009) 43–56.

90–91 SHOULD I BUY GRAFTED PLANTS?
K. Mudge et al., "A History of Grafting" *Horticultural Reviews*, 35 (2009). Alex Wilkins, "Near impossible plant-growing technique could revolutionise farming", *New Scientist*, 22 December 2021.

94 ARE SOME PLANTS BETTER GROWN TOGETHER?
R. P. Larkin, et al., "Rotation and cover crop effects on soil borne potato diseases, tuber yield, and soil microbial communities." *Plant Disease*, 94, no.12 (2010) 1491–1502. Jessica Walliser, *Plant Partners: Science-based Companion Planting Strategies for the Vegetable Garden*, Storey Publishing, 2021. "Push-pull cropping: fool the pests to feed the people",

Rothamsted Research [web article], https://www.rothamsted.ac.uk/push-pull-cropping.

95 SHOULD I ROTATE PLANTS EACH YEAR?
Mirza Hasanuzzaman, *Agronomic crops Vol. 1: Production Technologies*, Springer Nature, 2019. K. D. White, "Fallowing, crop rotation, and crop yields in Roman times" *Agricultural History*, 44, no.3 (1970) 281–290. "A guide to the nutritional requirements of crops", Adam Otter, IntelliDigest [web article], 2022, intellidigest.com/services research/a-guide-to-the- nutritional-requirements -of-crops/?doing_wp_ cron=1656602270.2639980316162109375000

100–101 HOW CAN I WORK OUT THE BEST SPOT FOR GROWING PLANTS INDOORS?
"Your plants get stressed when they're hot", Martha Proctor, University of California [web article], ucanr.edu/sites/MarinMG/files/152980.pdf. "Artificial lighting for indoor plants", Royal Horticultural Society [web article], rhs.org.uk/plants/types/houseplants/artificial-lighting.

104–105 ARE ALL ROOTS THE SAME?
Y. Liu et al., "A new method to optimize root order classification based on the diameter interval of fine root", *Sci. Rep.*, 8 (2018) 2960. Maire Holz et al., "Root hairs increase rhizosphere extension and carbon input to soil", *Annals of Botany*, 121, no. 1 (2018) 61–69.

106–107 CAN I SIMPLY DIG UP AND MOVE PLANTS?
P. Alvarez-Uria, C. Korner, "Low temperature limits of root growth in deciduous and evergreen temperate tree species", *Functional Ecology* 21, no.2 (2007) 211–218. "Transplanting – should you reduce top growth?", Robert Pavlis, Garden Myths [web article], www.gardenmyths.com/transplanting-should-you-reduce-top-growth/. "Trees and shrubs: moving plants", Royal Horticultural Society [web article], rhs.org.uk/plants/types/trees/moving-trees-shrubs.

108 DO ALL PLANTS GET THEIR NUTRIENTS FROM SOIL?
Gergo Palfalvi et al., "Genomes of the Venus flytrap and close relatives unveil the roots of plant carnivory", *Current Biology*, 30, no.12 (2020) 2312–2320. A. M. Ellison et al., "Energetics and the evolution of carnivorous plants—Darwin's 'most wonderful plants in the world'", *Journal of Experimental Botany*, 60, no.1 (2009) 19–42.

114–115 WHAT IS THE BEST WAY TO WATER PLANTS?
C. Brouwer, K. Prins, M. Heibloem, Irrigation water management: training manual no.5: irrigation methods, Annex 2 Infiltration rate and infiltration test, Food and Agriculture Organization of the United Nations, 1985. D. Dietrich et al., "Root hydrotropism is controlled via a cortex-specific growth mechanism" *Nature Plants*, 3 (2017). S. Nxawe et al., "Effect of regulated irrigation water temperature on hydroponics production of Spinach (*Spinacia oleracea* L.)", *African Journal of Agricultural Research*, 4, no.12 (2009) 1442–1446. Andy McMurray, "Effects of water temperature on Easter lilies", *North Carolina Flower Growers' Bulletin*, 22, no.2, (1978). Jay W. Pscheidt, *Flourine toxicity in plants*, Pacific Northwest Pest Management Handbooks, pnwhandbooks.org/plantdisease/pathogen-articles/nonpathogenic-phenomena/fluorine-toxicity-plants.

116 HOW DO PLANTS COPE WITH WET CONDITIONS?
Samuel Taylor Coleridge, The Rime of the Ancient Mariner (1834). Pan Jiawei et al., "Mechanisms of waterlogging tolerance in plants: research progress and prospects", *Frontiers in Plant Science*, 11 (2021).

117 HOW DO PLANTS COPE WITH DROUGHT?
Cruz de Carvalho, Maria Helena, "Drought stress and reactive oxygen species: production, scavenging and signaling", *Plant signaling & behavior*, 3, no.3 (2008) 156–65. El Khoumsi Wafae et al., "Integration of groundwater resources in water management for better sustainability of the oasis ecosystems – case study of Tafilalet Plain, Morocco", *3rd World Irrigation Forum* (2019).

120–121 WHAT'S THE BEST WAY TO RAISE HUMIDITY FOR INDOOR PLANTS?
"Tropical rainforest biomes", Khan Academy [web article], khanacademy.org/science/biology/ecology/biogeography/. Richard Slávik, Miroslav Cekon, "Hygrothermal loads of building components in bathroom of dwellings", *Advanced Materials Research*, 1041 (2014) 269–272. "Increasing humidity for indoor plants: what works and what doesn't", Robert Pavlis, Garden Myths [web article], gardenmyths.com/increasing-humidity-indoor-plants/.

122–123 WHICH NUTRIENTS ARE ESSENTIAL FOR HEALTHY PLANTS?
Janet I. Sprent, Euan K. James, "Legume evolution: where do nodules and mycorrhizas fit in?" *Plant Physiology*, 144, no.2 (2007) 575–81.

126–127 WHAT'S THE SECRET TO A HEALTHY, ATTRACTIVE LAWN?
"Different photosynthesis rates show the grass really is greener sometimes", Ian Chant, The Mary Sue [web article], 2012, themarysue.com/greener-grass/. "Grass holds the secret to more efficient crops?" Belinda Smith, Cosmos [web article], 2016, cosmosmagazine.com/science/biology/does-grass-hold-the-secret-to-more-efficient-crops/. H. Chen et al., "The extent and pathways of nitrogen loss in turfgrass systems: age impacts.", The Science of the total environment, 637–638, (2018) 746–757. "How to have and care for a healthy lawn: top 7 non-negotiables", Joe Lamp'l [web article], 2018, joegardener.com/podcast/healthy-lawn-care/. University of Hertfordshire Pesticide Properties Database, sitem.herts.ac.uk/aeru/ppdb/en/Reports/431.htm

128–129 HOW DO PLANTS CLIMB?
"English ivy's climbing secrets revealed by scientists", Jody Bourton, Earth News, 28 May 2010, news.bbc.co.uk/earth/hi/earth_news/newsid_8701000/8701358.stm.

140–141 WHICH PLANTS ARE BEST FOR POLLINATORS?
"Plummeting insect numbers 'threaten collapse of nature'", Damian Carrington, *The Guardian* [web article], 10 Feb 2019. "The assessment report of the Intergovernmental Science-Policy Platform on biodiversity and ecosystem services on pollinators, pollination and food production", IPBES (2016), S.G. Potts, V. L. Imperatriz-Fonseca, and H. T. Ngo (eds). Secretariat of the Intergovernmental

Science-Policy Platform on Biodiversity and Ecosystem Services. James C. Rodger, et al., "Widespread vulnerability of flowering plant seed production to pollinator declines", *Science Advances*, 7, no.42 (2021).

142–143 HOW DO I ENCOURAGE HOUSEPLANTS TO FLOWER?
S. N. Freytes, et al., "Regulation of flowering time: when and where?", *Curr. Opin. Plant Biol.*, 63 (2021). F. Andrés et al., "Analysis of photoperiod sensitivity sheds light on the role of phytochromes in photoperiodic flowering in rice", *Plant Physiol.*, 151, no.2 (2009) 681–690. Yin-Tung Wang, "Impact of a high phosphorus fertilizer and timing of termination of fertilization on flowering of a hybrid moth orchid", *HortScience*, 35, no.1 (2000).

144 SHOULD I GROW MY OWN CUT FLOWERS?
Jeannette Haviland-Jones et al., "An environmental approach to positive emotions: flowers", *Evolutionary Psychology*, 3 (2005). H. Ikei et al., "The physiological and psychological relaxing effects of viewing rose flowers in office workers", *Journal of Physiological Anthropology*, 33, no.1 (2014).

146–147 WHAT IS BOLTING AND HOW DO I PREVENT IT?
"Why do Greens Bolt?", Megan Haney, Fine Gardening, Issue 164 [web article], finegardening.com/project-guides/fruits-and-vegetables/why-do-greens-bolt

149 WHY DOES MY TREE PRODUCE FRUITS EVERY OTHER YEAR?
"Understanding crop load and growth regulator effects on biennial bearing in apple trees" Christopher Gottschalk et al., Michigan State University [web article], canr.msu.edu/uploads/files/16_treefruit_Gottschalk.pdf.

154–155 WHAT MAKES LEAVES CHANGE COLOUR AND WHY DO THEY THEN DROP?
Ines Pena-Novas, Marco Archetti, "A test of the photoprotection hypothesis for the evolution of autumn colours: chlorophyll resorption, not anthocyanin production, is correlated with nitrogen translocation", *Journal of Evol. Biology*, 34, no.9 (2021) 1423–1431. K. S. Gould, "Nature's Swiss army knife: the diverse protective roles of anthocyanins in leaves", *Journal of Biomed. Biotech.*, 5 (2004) 314–320.

158–159 HOW DO I CARE FOR MY GARDEN OVER WINTER?
Johannes Heinze et al., "Soil temperature modifies effects of soil biota on plant growth", *Journal of Plant Ecology*, 10, no.5 (2017) 808–821.

163 WHY DO MY HOUSEPLANTS DIE OVER WINTER?
Alexander S. Lukatin, "Chilling injury in chilling-sensitive plants: a review", *Žemdirbystė Agriculture*, 99, 2 (2012) 111–124.

166–167 WHAT HAPPENS AFTER I MAKE A PRUNING CUT?
R. P. Baayen et al., "Compartmentalization of decay in carnations", *Phytopathology*, 86, no. 10 (1996).

170–171 DOES IT MATTER WHERE I MAKE A PRUNING CUT?
L. Chalker-Scott, A. J. Downer, "Myth busting for extension educators: reviewing the literature on pruning woody plants",

Journal of the NACAA, 14, no. 2, (2021).

172–173 SHOULD I TRAIN A FRUIT TREE AGAINST A WALL OR FENCE?
"Fruit Tree Pruning – Basic Principles", Robert Crassweller, PennState Extension [web article], 2017, extension.psu.edu/fruit-tree-pruning-basic-principles. Nikolaos Koutinas et al., "Flower Induction and Flower Bud Development in Apple and Sweet Cherry", *Biotech. & Biotechnological Equipment*, 24, no.1 (2010) 1549–1558.

176–177 CAN I COLLECT MY OWN SEEDS TO RAISE NEW PLANTS?
S. Takayama et al., "Direct ligand-receptor complex interaction controls brassica self-incompatibility", *Nature*, 4, no. 413 (2001). H. Shimosato et al., "Characterization of the SP11/SCR high-affinity binding site involved in self/non-self recognition in brassica self-incompatibility", *Plant Cell*, 19, no.1 (2007) 107–117. "Types of plants that can't self-pollinate", Lori Norris, SFGate [web article], homeguides.sfgate.com/types-plants-cant-selfpollinate-80879.html.

178 WHAT'S THE BEST WAY TO STORE SEEDS?
"Successful seed storage at home", Kevin McGinn, National Botanic Garden Wales [web article], (2020), botanicgarden.wales/2020/08/successful-seed-storage-at-home/. "Seed: collecting and storing", Royal Horticultural Society [web article], rhs.org.uk/propagation/seed-collecting-storing.

179 HOW LONG DO SEEDS REMAIN VIABLE?
Janine Wiebach et al., "Age-dependent loss of seed viability is associated with increased lipid oxidation and hydrolysis", *Plant, Cell, and Environment*, 43, no. 2 (2020). Loïc Rajjou et al., "Seed longevity: survival and maintenance of high germination ability of dry seeds", *Comptes Rendus Biologies*, 331, no.10 (2008) 796–805. "How does the age of a seed affect its ability to germinate?", Laura Reynolds, SFGate [web article] homeguides.sfgate.com/age-seed-affect-its-ability-germinate-69423.html. "How long do seeds last?", Aaron von Frank, Grow Journey [web article], growjourney.com/long-seeds-last-seed-longevity-storage-guide. "Common poppy", Garden Organic [web article], gardenorganic.org.uk/weeds.

182–183 IF I SPLIT UP A PLANT WILL IT DIE?
José León et al., "Wound signalling in plants", *Journal of Experimental Botany*, 52, no. 354 (2001) 1–9.

184 CAN I GROW ANYTHING FROM CUTTINGS?
A. J. Koo, G. A. Howe, "The wound hormone jasmonate", *Phytochemistry* 70, no.13–14 (2009) 1571–1580.

188–189 WHAT'S COMPOST AND HOW DOES IT FORM?
"Understanding soil microbes and nutrient recycling", James J. Hoorman, Rafiq Islam, Ohio State University Extension, 2010, ohioline.osu.edu/factsheet/SAG-16.

192 SHOULD I AVOID PUTTING WEEDS AND DISEASED LEAVES IN COMPOST?
Ruth M. Dahlquist et al., "Time and temperature requirements for weed seed thermal death", *Weed Science*, 55 (2007) 619–625.

193 WHY DO I NEED TO LET MANURE ROT DOWN?
X. Jiang et al., "The role of animal manure in the contamination of fresh food", *Advances in Microbial Food Safety*, 2015, 312–350.

194–195 WHAT SORT OF DISEASES DO PLANTS GET?
"Plant pathology guidelines for master gardeners", Richard Reid [web article], erec.ifas.ufl.edu/plant_pathology_guidelines/ module_05.shtml. V. A. Robert, A. Casadevall, "Vertebrate endothermy restricts most fungi as potential pathogens", *J. Infect. Dis.*, 200, no.10 (2009) 1623–1626.

196–197 HOW CAN I STOP MY PLANTS GETTING DISEASED?
Kim E. Hammond-Kosack, Jonathan D. G. Jones, "Plant Disease Resistance Genes", *Annu. Rev. Plant Physiol. Plant Mol. Biol.*, 48 (1997) 575–607. B.C. Freeman, G.A. Beattie, "An overview of plant defenses against pathogens and herbivores", *The Plant Health Instructor*, (2008). "Plant bacteria thrive in wet weather", Neha Jain, Science Connected Magazine [web article], 2022, magazine. scienceconnected.org/2022/03/plant-bacteria-thrive-wet-weather/.

200–201 WHAT'S THE BEST WAY TO PREVENT PEST ATTACK?
E. J. Andersen et al., "Disease resistance mechanisms in plants", *Genes*, 9, no.7 (2018) 339. Carolyn Mitchell et al., "Plant defense against herbivorous pests: exploiting resistance and tolerance traits for sustainable crop protection", *Front. Plant Sci.*, 7 (2016). "Why insect pests love monocultures, and how plant diversity could change that", Science Daily [web article], 2016, sciencedaily.com/ releases/2016/10/ 161012134054.htm. S. Pascual et al., "Effects of processed kaolin on pests and non-target arthropods in a Spanish olive grove", *J. Pest Sci.*, 83 (2010) 121–133. "Should I buy ladybugs for the garden?", Robert Pavlis, Garden Myths [web article], gardenmyths.com/buy-ladybugs-garden/.

202–203 HOW CAN I PREVENT SLUGS AND SNAILS DAMAGING PLANTS?
Planet friendly: RHS no longer to class slugs and snails as pests. *The Guardian* 4 March 2022. https://www.theguardian.com/ environment/2022/mar/04/planet-friendly-rhs-to-no-longer- class-slugs-and-snails-as-pests. "New study disproves myths to get rid of slugs", N. Mason, Pro Landscaper [web article], 2018, prolandscapermagazine.com/myths-deterring-slugs/. Azlina Mat Saad et al., "Metaldehyde toxicity: a brief on three different perspectives", *Journal of Civil Engineering, Science and Technology*, 8, no.2 (2017). "Less toxic iron phosphate slug bait proves effective", Glenn Fisher, Oregon State Uni Extension Service [web article], 2008, extension.oregonstate.edu/news/less-toxic-iron-phosphate- slug-bait-proves-effective.

208–211 GARDENING MYTHS
"Are gardeners wrong to put crocks in pots?", Tom de Castella, BBC News [web article] bbc.co.uk/news/blogs-magazine-monitor- 27126160. "How Many Plants Would It Take to Produce Enough Oxygen for One Person?" Candide Gardening [web article], medium. com/@candidegardening/how-many-plants-would-it-take-to- produce-enough-oxygen-for-one-person-7312743ed70b.

Art Editor Alison Gardner
Project Editor Jo Whittingham
Gardening Consultants Mike Grant, Phil Clayton
Illustrator Sally Caulwell
Photographers Gary Ombler, Ian Gilmour
Consultant Gardening Publisher Chris Young

FOR DK
Senior Designer Louise Brigenshaw
Senior Editor Alastair Laing
DTP and Design Coordinator Heather Blagden
Production Editor David Almond
Production Controller Rebecca Parton
Jacket Designer Amy Cox
Jacket Coordinator Jasmin Lennie
Design Manager Marianne Markham
Managing Editor Ruth O'Rourke
Art Director Maxine Pedliham
Publishing Director Katie Cowan

First published in Great Britain in 2023 by
Dorling Kindersley Limited
DK, One Embassy Gardens, 8 Viaduct Gardens,
London, SW11 7BW

The authorised representative in the EEA is
Dorling Kindersley Verlag GmbH. Arnulfstr. 124,
80636 Munich, Germany

Text copyright © Dr Stuart Farrimond 2023
Copyright © 2023 Dorling Kindersley Limited
A Penguin Random House Company
10 9 8 7 6 5 4 3 2
007-328771-Mar/2023

A CIP catalogue record for this book
is available from the British Library.
ISBN: 978-0-2415-5925-3

Printed and bound in China

For the curious
www.dk.com

MIX
Paper | Supporting responsible forestry
FSC™ C018179

This book was made with Forest Stewardship Council ™ certified paper – one small step in DK's commitment to a sustainable future. For more information go to www.dk.com/our-green-pledge

ACKNOWLEDGEMENTS

Like the web of life upon which each plant depends, this book would not exist without the unseen efforts of many people. First and foremost, I am indebted to editor Jo Whittingham, who has been an unflappable source of support, encouragement, and knowledge. This book is unique for twining together practical know-how with the latest scientific research – and when my hunt for answers led me into a cul-de-sac, plant science guru and all-round nice guy, Mike Grant answered questions and provided direction.

I am once again staggered by the creativity and skill of designer Alison Gardner, who has devised the book's beautiful diagrams, sometimes just from my scrawlings on scrap paper. Questions about soil science were answered by Josef Carey, "plant physician, photosynthesis manager, soil rehabilitator" and CEO of Swedish organisation 59 degrees, which helps gardeners grow a better future by improving the life within their soil. New Zealand's leading professional breeder of ornamental plants, Dr Keith Hammett, was gracious enough to answer questions on this dark art. Dr Stuart Tustin helped me through the tangled thicket of pruning by sharing his 30-plus years of research in fruit tree growing methods.

Thanks also to Dawn Henderson, Ruth O'Rourke, Alastair Laing, and the DK team, for sharing my vision of a book for everyone written by an outsider in the gardening world. I am immeasurably thankful to my wonderful and ever-supportive wife, Grace, who has both endured the imposition of my growing experiments in "her garden" and tolerated me giving her gardening advice that she already knew! Thanks also to Jonny Pegg – a dependable cheerleader for me and my work, and always on hand when there is a bump in the road. Finally, apologies go to those who have shared time and expertise but whose contributions have fallen from my memory.

PUBLISHER'S ACKNOWLEDGEMENTS

Thanks to Marie Lorimer for indexing; Alice McKeever for proofreading; Steven Marsden and Eloise Grohs for design assistance; Nityanand Kumar for repro work; Adam Brackenbury for image retouching; and Aditya Katyal for picture research.

ABOUT THE AUTHOR

Dr Stuart Farrimond is a medical doctor turned science communicator and award-winning writer. He is author of *The Science of Cooking* (2017) and *The Science of Spice* (2018), and *The Sunday Times* bestseller, *The Science of Living* (2021) (sold as *Live Your Best Life* in North America). He makes regular appearances on TV, radio, and at public events, and his writing appears in national and international publications, including *The Independent*, *The Daily Mail*, and *New Scientist*. A qualified teacher and a former University of Cambridge tutor, Stuart's passion is to explain the science behind the everyday. Since 2017, Dr Stu has been the food scientist for BBC's much-loved show Inside the Factory, hosted by Gregg Wallace and Cherry Healey.

Stuart is as a cancer survivor and an ambassador for brain tumour patients. He has raised over £10,000 for brain tumour charities and lobbies for improved brain tumour research funding. A keen cyclist and grower of pumpkins, Stuart lives in the town of Trowbridge, Wiltshire, UK with his wife, Grace, Winston the dog, and far too many plants.